David F. Batten · Roland Thord (Eds.)

Transportation for the Future

With Contributions by
Å. E. Andersson, D. F. Batten, D. E. Boyce, G. C. Eads
S. Ericsson, R. L. French, H. Fujii, W. L. Garrison
J. Grevsmähl, T. Karlsson, K. Krell, N. Nakicenovic
D. Näsman, C. Nicolin, H. St. Seidenfus, J. van Stappen
U. Strömquist, R. Thord

With 58 Figures

Springer-Verlag Berlin Heidelberg New York
London Paris Tokyo Hong Kong

Professor Dr. David F. Batten, Centre for Regional Science Research and Department of Economics, University of Umeå, S-901 87 Umeå, Sweden

Dr. Roland Thord, Temaplan AB, Applied Systems Analysis for Industry and Government, Sandhamnsgatan 81, S-115 28 Stockholm, Sweden

ISBN-13: 978-3-642-74868-4 e-ISBN-13: 978-3-642-74866-0
DOI: 10.1007/978-3-642-74866-0

2142/7130-543210

PREFACE

During the seventies and eighties, we have witnessed some profound changes in those industries associated with the transportation of goods and people. The quickening pace of technological change, with its expanding emphasis on telecommunications, knowledge handling capacity, and air transportation, has increased the discrete character of the world economy. This means that the *network structure* of global development patterns is of growing importance, with metropolitan centres as key nodes and rapid transportation paths as key links.

In this evolutionary situation, changes in the preferred mix of transportation modes are inevitable. The faster and more direct modes will be favoured, individually and in combination. Changes in transportation logistics are also of growing significance. As one manager of a Swedish systems-oriented corporation reported in a recent interview: "Our aim is to extend our conveyor belt across all the highways of Europe". This also implies an increased interest from systems-oriented corporations in public transportation policies and the provision of infrastructure.

This volume contains a collection of papers presented at the First International Conference on *Transportation for the Future*, held at the Hotel Skogshöjd in Södertälje, Sweden from 24-25th May 1988. These papers addressed the above-mentioned themes from several disciplinary perspectives: engineering, transportation logistics, communication science, urban and regional planning, systems analysis and economics, to name a few. The aim of the conference was to bring together about twenty industry leaders and prominent scientists from different parts of the world in order to stimulate an environment of spontaneous exchange - in which common threads and points of difference could be permitted to "self-organize" into a reasonably holistic view of how the transportation industry might grow and change in a long-term perspective.

The Conference was initiated by the Swedish Board for Technical Development, the Swedish National Road Administration, and the Community of Södertälje, who also served as joint hosts for its duration. Practical arrangements were undertaken by an Organizing Committee consisting of staff from the Södertälje Business and Technology Centre and from Temaplan AB. The Scania Division of Saab-Scania AB kindly arranged the Conference Banquet. Professor Åke E. Andersson assisted with the preliminary planning of the programme and Professor Carl-Olof Ternryd delivered the opening speech. An edited version of his speech appears in the Foreword to this volume.

As well as thanking each of the above organizations and individuals for their support and cooperation, we wish to acknowledge the assistance provided by the Swedish Transport Research Board (TFB) and the Centre for Regional Science Research (CERUM) towards the production of this volume. We would also like to thank each of the contributing authors for their valuable ideas and suggestions both during and after the Conference, and Iréne Sundwall (Temaplan AB) for her skilful preparation of the complete manuscript.

David Batten and Roland Thord
Stockholm, April 1989

FOREWORD

Three years ago we arranged in Stockholm (at the Royal Academy of Engineering Sciences) a series of discussions on the interdependencies between R and D, Industrial Development and Security Policy in Sweden. There were more than six hundred managers and research leaders from industry, professors from academia and representatives of various public authorities participating in these seminars. The conclusions and recommendations stated that the global internationalization of industry during recent years will continue and pervade practically all of the important sectors of our Society.

The consequences of internationalization for Sweden are more obvious than for some other countries. This is due to the fact that Sweden is a highly industrialized country with an advanced technology and a limited homemarket.

After a period of economic and industrial stagnation during the latter part of the 70s, Swedish industry has recovered well during the 80s. It has gone through a swift transition during a short period of time. Practically every company and every economic sector in Sweden has undergone reorganization or rationalization. The drive to improve productivity, quality and the efficiency of capital expenditure has been relentless. New technologies have been introduced at a rapid pace. Modern and capital-effective philosophy concerning inventory and logistics has been developed and introduced, having an important influence on economic performance. Small inventory - which means a rapid capital flow and low capital costs - can become effective only if the transportation and logistical system is so well developed that it embraces the *JUST-IN-TIME* concept. An effective infrastructure is needed nationally as well as for international contacts.

Industrial research

A small country like Sweden is strongly dependent on other countries for the development of its industry and infrastructure. We represent a mere one percent of all research in the world and we are therefore obliged to rely on research results from other nations. In order to be interesting as partners for the exchange of research results we have to be excellent in some areas. We must find some interesting niches in which we can excel.

But opinion on preferable industrial developments changes rapidly. A few years ago an international consulting group maintained that the *Swedish pulp and paper industries* were doomed, but these industries have recently demonstrated an unusual capacity for innovative thinking and renewal. The Swedish machinery manufacturers and pulp producers have combined their resources and utilized new technological advances to produce new types of mechanical pulp and tissue papers. Their researchers have recently presented new techniques for obtaining three times the amount of high quality printing paper from a given quantity of wood at lower costs. They use biochemically modified starch and simple fillers. Today 20 percent of Swedish exports come from timber.

Another branch that was supposedly doomed a few years ago was the *Swedish car industry*. According to some European specialists, there would only be space for three or four car manufacturers in Europe by the end of the 80s, and the Swedish makers were not among them. Today the Swedish automotive giants *VOLVO* and *SAAB-SCANIA* sell more cars, buses and trucks than they can produce. They have found the correct niches and they have developed an effective production and marketing organization.

These few examples from Swedish industry demonstrate clearly that investments in education, research and development are of the utmost importance for the economic development of a country like Sweden. The entrepreneurial management of technology and the companies using it is another very important factor for the successful development of industry and society at large. Great attention is therefore paid to management training on different levels in our country.

The Swedish production costs are among the highest in the world, partly due to our high labour costs. To compensate for this fact and to compete successfully on the international market, we must be able to confront our rivals with superior product quality and more cost-effective production methods. It is no coincidence that Sweden has more robots per capita than any other nation. *ASEA* is the world´s largest manufacturer of robots. These robots have a great impact on product quality. Effective coordination between man and machine is considered one strength of our method of applying new technology.

Looking forward

A high level of technology within specialized niches is increasingly important for industrial competitiveness on the international market. Close cooperation between Swedish and foreign enterprises has been organized within production, marketing and distribution. Swedish foreign trade and Swedish industries are well established within Western Europe and the USA. The importance of raw materials is decreasing and the products of engineering, pharmaceutical, chemical and biochemical industries are increasing. Therefore, Sweden s future competitiveness on the world market lies in those industrial fields and market segments characterized by high standards of quality and where services are integrated closely with the products.

As in many other highly industrialized countries, we are entering what has been called an *Information Society*. The need for rapid communications is growing. Systems for the daily transportation of people to and from their workplaces will become even more important in the future. Today many of the world's major cities suffer from enormous problems associated with automobile traffic in the suburbs and the central areas. From an environmental standpoint, we must find rational solutions to these problems.

There is also a similar increase in the need for cheap and reliable energy. Environmental requirements are becoming stricter. This fact emphasizes the need for equipment that can meet such demands. Both transport and energy supply systems are part of our nation's infrastructure. To develop and maintain a smoothly functioning infrastructure and a skilled and competitive industrial sector is essential for the future of Sweden. We believe that there exist these possibilities in our country.

Sweden's ability to maintain and to strengthen her position as a successful industrial nation is entirely dependent on how successful we are in various export markets. Closer cooperation within the framework of the European Community is a necessity. Continued internationalization of Swedish industry is a necessity for further expansion. This will take place in the form of newly established operations in strategic markets, acquisitions, mergers and cooperation agreements with foreign companies.

Development and maintenance of physical infrastructure has a decisive impact on the economic and social development of a nation. To move people and goods in a cost-effective way is becoming more and more important. There must be possibilities to arrange an effective equilibrium between the different transportation modes, and our transportation research must be channeled towards these requirements. There must be more space for rational solutions than for nostalgic thinking. Technological advances during recent years, and forecasts for the near future, present opportunities that should be utilized in order to promote transportation effectiveness and safety.

This First International Conference on *TRANSPORTATION FOR THE FUTURE* is of the greatest importance for a better understanding of the role which a robust infrastructure can play in the future development of society and its industry. I wish you all every success in your work here at Södertälje, and I am quite convinced that some excellent results will emerge from your discussions.

Carl-Olof Ternryd
Chairman, The Association of Swedish Automobile Manufactures and Wholesalers

CONTENTS

Chapter 1

INTRODUCTION

David F. Batten and Roland Thord
University of Umeå and TEMAPLAN AB, Sweden

1.1 PREAMBLE

Studies in various parts of the world have demonstrated that the demand for transport of goods and passengers usually increases more quickly than the GDP per capita. In the developed economies, transport demand grows annually by some 20-25 percent more than the growth of GDP per capita; whereas in the developing economies the growth rate of transport surpasses that of GDP per capita by up to 50 percent. European examples of this growth differential for freight and passenger transport are given in Table 1.1.

Table 1.1 Transport Demand and GDP per Capita in the ECMT Countries

ECMT Average	Year 1975	1980	1984	Average Growth Rate Per Annum 1975-1984
GDP per capita at constant prices (U.S. dollars)	4467	5072	5185	1.67% p.a.
Passenger transport* (kms per capita)	6000	6822	7179	2.01% p.a.
Freight transport* (tonne-kms per capita)	2106	2449	2502	1.93% p.a.

* Surface modes of transport only

Source: Adapted from the European Conference of Ministers of Transport (1988).

This growing emphasis on movement is also evident in the changing share of household expenditure devoted to transportation. People in lesser developed societies commit less than five percent of their household incomes to travel and communication, whereas in more developed societies the average figure per person is closer to fifteen percent of disposable income and one hour of travel per day (see Zahavi, 1979; also Chapter 5 in this volume). By the year 2000 the share of income devoted to transport may reach 20 percent.

Both freight and passenger transport attract a larger share of economic growth as average incomes rise and the cost/km of newer forms of transportation falls in real terms, because people and firms allocate increasing shares of their travel time to these newer and faster transport modes. Looking into the future, it is pertinent to ask whether this expansive trend can continue indefinitely?

One provocative view of the future is depicted in Figure 1.1. Bernal's vision of changes in occupational structure foresees a relatively rapid structural adjustment evolving towards a knowledge-oriented society in which science and engineering steer the economy, serviced predominantly by transport, communication and other service occupations. This is a society where personal contacts, systems-oriented firms and logistical networks play a catalytic role (see Andersson, 1985), and in which the need for further growth in transport services is fundamental.

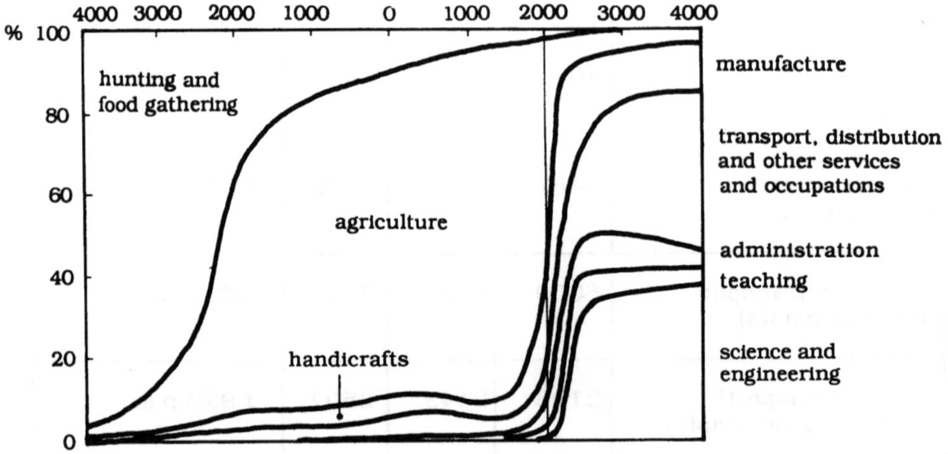

Figure 1.1 Changes in occupation in the past and the future.
Source: Bernal (1958).

Two questions spring to mind immediately. Is the growth in transport services associated with Bernal's society of the future technically feasible? If so, is it socially and politically preferable?

In many post-industrial societies, the willingness to invest more heavily in certain forms of transport infrastructure has waned appreciably (see Figure 1.2). Road investment has declined in real terms, reflecting the lack of harmony between cars and people in densely populated cities, environmental pressures to reduce pollution levels, and a perceived need to liberate cities from their ugly pavement "jungles". Yet road transport continues to service the lion's share of both the goods and the passenger markets in North America and Europe. Thus there are good reasons to believe that air transport will take on a greater share of the movement market between cities as the tendency to discourage more automobiles increases (Figure 1.3).

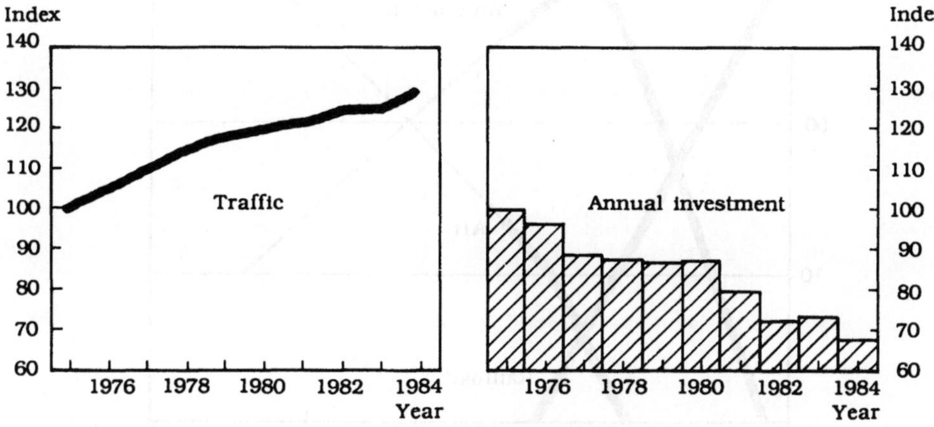

Figure 1.2 Trends in Annual Road Investment and Road Traffic within The ECMT Nations, 1975 = 100
Source: ECMT (1988)

Such an evolutionary pattern among modal options is in keeping with the observed transition of investments in transportation infrastructure, namely the sequence: canals → railways → paved roads → air networks. There are many similar examples of multiple competition and gradual replacement within the gamut of individual transport technologies. Logistical substitution models have been widely applied in order to explain the key changes underway and to forecast accurately well into the future. This competitive perspective on transportation futures is explored more intensively in Part II of this volume.

1.2 EXPLORING THE THIRD DIMENSION

With the advent of the aeroplane, transportation systems finally broke out of the restrictive spatial domain imposed by surface movement. This third dimension offers vast scope for personal and freight transportation in the future. The impact of aerospace technology on the international economy is amply demonstrated within the USA, where it has the longest track record. During the latter half of the eighties, the U.S. aerospace industry has generated its greatest level of export sales and a substantial trade surplus in its own right. This continues to have a stabilizing effect on the U.S. economy at a time when that nation is experiencing the worst overall trade deficit in its history.

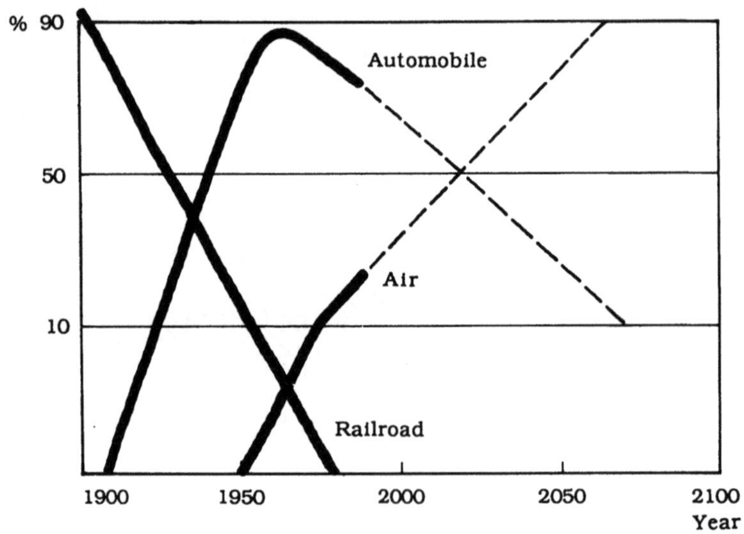

Figure 1.3 A Forecast of Intercity Personal Transport in the USA by Mode. Source: Andersson et. al., (1987)

The pace of developments in airborne transport has been rapid. According to statistics from the International Civil Aviation Organization (ICAO), between 1970 and 1986 the total passenger-kilometres on scheduled air services throughout the world trebled, the average passenger load factor improved from 55% to 65%, and the ton-kilometers of freight carried almost quadrupled. The total ton-kilometers performed by world air traffic system has been growing steadily and is expected to reach the end of a Kondratiev cycle of growth around 1995 (Marchetti, 1987).

Assuming that private travellers and firms moving goods both tend to allocate time and money to different modes of transport so as to maximize range, faster modes must be favoured eventually. Since the aeroplane is the fastest mode currently available, its real cost is declining slowly, and average disposable incomes are still on the increase, air transport ought to be the fastest-growing branch of all. Poorer people walk or cycle, richer ones fly, and those in between drive or take public transport. The evolutionary pattern portrayed in Figure 1.3 gains greater credibility once it is recognized that the real value of time is positively correlated with rising incomes.

Are there any new competitors looming upon the horizon? Confining our options to "third dimension" possibilities, two candidates spring to mind. Magnetic levitation trains (Maglev) might achieve the necessary speed range (>500 km/hour) to represent a significant improvement for intercity travel. Vacuum tunnels may also satisfy similar speed criteria, although the locations of feasible trajectories may be limited.

1.3 COMBINED TRANSPORTATION SYSTEMS

Although the choice of transport mode is often regarded as a question of seeking the dominant mode (e.g. Figure 1.3), in reality transportation has always consisted of a combination of solutions. In the early days, canal transport was common for long-haul segments, whereas horses were used for the actual pickup and delivery segments. During the period of industrialization in the 1800s and early 1900s, a combined system of ship (long haul) and rail (short haul) was adopted. Railways sometimes combined with other railways. The postwar era has seen the emergence of trucks in combination with ships, trains, and other trucks.

As the average value/weight ratio of goods increases, it becomes economical to utilize the faster modes (road, air) in combination with themselves. Truck combinations or trucks with "door-to-door" transport are becoming the most feasible solution for much of Europe. The SAS Cargo System is an example of the plane-truck combination, with the terminal at Arlanda serving as a principal node in such a combined system. Cologne is another key node in this system. Another example is the combined train-truck system between Scandinavia and mainland Europe, to be facilitated by the Scandinavian Link project (see Chapter 2).

The historical evolution of goods traffic, and the future spectrum of possibilities, can be understood in terms of the various combinations mentioned in Table 1.2.

Table 1.2 The Development of Freight Transport in terms of Kombi-Solutions.

	SHIP	**TRAIN**	**TRUCK**	**AEROPLANE**
SHIP	**The Canal Era**			
TRAIN	**Kombi-system for the North Atlantic´s industrial era**	**The Continent´s kombi-solution for the industrial era**		
TRUCK	**The predominant industrial kombi-system**	**The new kombi-system**	**The kombi-system for geographically decentralized industrialists**	
AEROPLANE	**An unlikely combination (except in crises)**	**Only for personal transport in densely-populated areas**	**Kombi-system for goods of very high freight value**	**Emergency situations and transport of VIP´s**

Source: Adapted from Andersson and Strömquist (1988)

1.4 OUTLINE OF THE BOOK

This book focuses predominantly on the future options for surface modes of transportation. In particular, many of the chapters acknowledge the ongoing transformation of production and distribution systems among firms striving to achieve just-in-time methods, economies of scope and an integrated approach to their economic activities.

Three visionary perspectives of future transportation and distribution systems appear in Part I. First, Curt Nicolin explores European goods transport over the next decade. He portrays the Scandinavian Link project as a timely example of infrastructure investments (in road, rail and fixed links) associated with the development of a kombi-system (truck-train) for transport between the Nordic countries and Central Europe. Speed, reliability and flexibility are key attributes of this integrated system. Next, George Eads reviews the future of the automobile. He argues that it will remain the primary passenger-carrying mode of transportation for some time to come, despite increasing concerns about energy conservation, pollution and traffic congestion. Although his perspective is based largely on the North American experience, his conclusions may extend to other parts of the world. The final paper in Part I is written by Åke E Andersson and Ulf Strömquist. They advocate that Europe's industrial and regional development patterns have been honed by the growth and change of logistical systems. In this context, the logistical infrastructure consists of the equipment and networks used for the transportation and distribution of people, commodities, information and knowledge. Andersson and Strömquist point to four logistical revolutions in the last millennium which have determined Europe's evolutionary path. The fourth and current logistical era stresses the importance of fast and flexible transportation modes, accumulated stocks and flows of knowledge, and the fully integrated production/distribution system.

Part II commences with a review of the evolution of transport systems over the centuries. Nebojsa Nakicenovic demonstrates that these systems evolve by steady processes of growth and senescence, with the dominant modes at any one time being subject to constant competitive pressures from the newer, faster and more reliable ones. David Batten also probes this type of technological change process, but from the debatable viewpoint of telecommunications as a potential substitute for travel. Arguments for and against the case of substitution growth are presented, before the author (weakly disguising his own bias) invites the reader to consider his or her own verdict. Finally, Bill Garrison portrays the transportation industry as a mature one seeking technologies which offer only marginal improvements, being limited in variety and efficacy. New innovative technologies may need instead to be associated with cleverly integrated systems of production and distribution.

Navigation technology is the theme of the first trio of papers addressing the future of surface transport. Robert French considers such technology to be the keystone of future onboard information systems which will guide the automobile or truck driver's decisionmaking concerning optimal routes and destinations. The minimum infrastructure needed to support such services is described. Haruki Fujii reviews the Comprehensive Automobile Control System (CACS) project undertaken by the Japanese Ministry of Trade and Industry (MITI) during the seventies. Major technological issues remain unresolved, even after the dynamic route guidance sytem (DRGS) was implemented experimentally on the road network in the Tokyo metropolitan area. This group of papers concludes with a chapter by David Boyce, discussing the potential contribution of transportation network theory and modeling to the development of real-time route guidance systems. Further unresolved research problems are identified and discussed.

There is a growing feeling that progress with much of the research into automated navigation aids has been slow and problematical. From the Prometheus program for improved road traffic in Europe, there is scant evidence of successful microelectronics adaptations which are suitable for widespread installation*). There seem to be many intentions but few inventions. One reason for this delay is the fundamental need for a true systems approach, combining microelectronics, information technology, road traffic and road infrastructure into a holistic solution which improves safety, efficiency and the environment. This point is stressed by a number of authors in the ensuing chapters.

Freight transport is the subject of two papers in Part III. Hellmuth St. Seidenfus discusses the future of truck transportation in Europe. Positing an optimistic economic outlook for western Europe to the year 2000, he perceives that road and air freight traffic will benefit from the trend towards more highly sophisticated goods and services. Transport's future role as part of a modern coordinated system of physical distribution is emphasized, with just-in-time principles growing in importance. Moving to the detailed level of truck technology, Stig Ericsson confirms Bill Garrison's view of a mature state of technical development. Reviewing truck development over the last forty years, he highlights the rapid improvements in the diesel-powered truck. The Scania developments are quoted as specific examples of truck design improvements.

Future trends in passenger transport and commuting behaviour are discussed by Karl Krell and Johannes Grevsmähl. Krell sees the growing application of information & communication technology as providing qualitative improvements but not significant replacements for passenger travel. Safer, more ecologically acceptable vehicles are needed to augment more flexible public transport solutions. The "park-and-ride" kombi solution is also discussed. Grevsmähl also sees private automobiles consolidating their position, through increased use by women, low income groups and the aged. Air transport is favoured for intercity travel, although rail systems may gain from novel uses of their established link infrastructure. Maglev is such a possibility.

The two final papers in the volume stress the growing importance of infrastructure - human and physical. Joseph von Stappen argues that the next generation of transport professionals will require new knowledge and skills to meet the challenge of future travel and distribution systems. A lead time of six to ten years is essential to ensure suitable competence. Dan Näsman stresses the very long run character of all planning for physical infrastructure investments, and the need to look well into the future. Investments in road networks have been decreasing steadily throughout Europe (see Figure 1.2), so much so that the annual

*) The Prometheus program is summarized in the Appendix.

depreciation incurred by the existing network now exceeds the annual investment. Although the problem of optimal road user charges for a satisfactory long-run equilibrium is a very difficult one among transport economists, some analytical tools are emerging (see, for example, Jansson 1984). A key topic for consideration is the serious discrepancy between the timing, location and level of investments in transportation infrastructure and the long run growth in demand for various transportation services. To help alleviate this worsening state of disequilibrium, our active contribution as transport researchers is urgently required.

REFERENCES

Andersson, Å. E. (1985). *Creativity and the Future of the Metropolis*, Stockholm: Prisma Press (In Swedish). (For an English summary of related ideas, see Andersson, Å. E. and D. F. Batten (1988) Creative nodes, logistical networks and the future of the metropolis, *Transportation* 14:281-293.)

Andersson, Å. E. et. al. (1987). *Future Transport in Sweden*. Ministry of Communication. Dsk 1987:16. Stockholm: Allmänna Förlaget. (In Swedish).

Andersson, Å. E. and U. Strömquist (1988). *The Future of the C-Society*, Stockholm: Prisma Press (In Swedish).

Bernal. J. D. (1958). *World Without War*, London: Routledge and Kegan Paul.

European Conference of Ministers of Transport (1988). *Investment in Transport Infrastructure in ECMT Countries*, Paris: ECMT.

Jansson, J. O. (1984). *Transport System Optimization and Pricing*, Chichester: John Wiley and Sons.

Marchetti, C (1987). *On Transport in Europe: The Last 50 Years and the Next 20*, paper presented at the First Forum on Future European Transport, Universität der Bundeswehr München.

Zahavi, Y. (1979). *The "UMOT" Project*. Report DOT-RSPA-DPB-20-79-3, prepared for the U.S. Department of Transportation and the Ministry of Transport, FRG. Springfield, Virginia: NTIS.

PART I

TRANSPORTATION AND DISTRIBUTION SYSTEMS: THREE VISIONARY PERSPECTIVES

Chapter 2

FUTURE GOODS TRANSPORT IN EUROPE

Curt Nicolin
ASEA Brown Boveri, Sweden and Switzerland

2.1 INTRODUCTION

In order to predict future developments in the field of international goods transport, it is necessary first to analyse some of the important factors which determine these developments. I shall concentrate on three such factors:

1. Changes in products and production systems.
2. Changes in the requirements of the world surrounding us concerning, for example, protection of the environment and road safety.
3. Changes in the possibilities offered by transport technology.

Our goods transport system of the future must, however, not just passively adapt to these developments. We also have a number of specific goals to fulfill. In Scandinavia in particular, these goals concern our ability to uphold our competitive position on the European as well as the overseas markets.

The picture I shall draw of future goods transport is, therefore, the result of a combination of effects brought to the transport field from the outside and of our own efforts to organize goods transport in such a way as to fulfill the requirements of industry.

2.2 EXTERNAL FACTORS

Let me begin by looking at the factors external to the transport fields as such. The most important ones are without doubt, changes in products and production systems. There is today hardly a design office anywhere in the world for which reduced product weight is not an important objective. From a transport point of view, this means, of course, that product value per kg is steadily increasing. At the same time, production systems reflect this development by eliminating costly storage and by sharply reducing lead times, the length of production series and finally the average size of transport consignments.

This indicates the very essential role that transport systems must play in a positive interaction with production systems. High quality goods transport is a sine qua non for efficient production systems in industry and thus for economic growth and expansion. Let me therefore briefly mention some of the important quality factors in future goods transport:

1. High punctuality/reliability
2. High frequency
3. High speed point-to-point
4. Sufficient capacity/flexibility
5. Low risk of goods damage

In particular, I would like to point to the importance of punctuality and reliability. As storage is more or less eliminated in the logistical systems of industry, uncertainty in the transport system entails extremely heavy costs in terms of delays in production and/or deliveries.

Frequency also plays a decisive role. As already mentioned, average consignment sizes will be reduced, logically leading to the need for greater frequency. For a given transport mode, frequency is also of the utmost importance. It is not unusual, for example, that a doubling in frequency leads to a tripling in transported volumes.

It goes without saying, of course, that costs are also of great importance. But here the focus should be on total costs in the production and distribution system as a whole, rather than on transport costs in isolation. We shall, therefore, see a trend towards high-performance and high-quality transport at higher costs, but leading to lower total costs and greater efficiency.

The second factor, which must be given increasing attention, is changes in the set of requirements, rules and regulations governing all economic activity, including transport. Let me mention in particular the problem of protecting the environment, which has come to play an increasing role in our industrialized countries. Road transport has - rightly or not - been blamed for contributing to air pollution, and we must expect increased restrictions on this transport form in the future.

Railways are generally found to be less polluting, at least if they run on electric power which is not generated by fossile fuels. In my view, however, changes in the transport pattern, e.g. from road to rail, should not be effected by restrictions, taxes or other negative instruments. Efforts should rather be made to increase the attractiveness, in terms of quality and price, of those means of transport one wishes to favour.

I find it of great importance that governments give all modes of goods transport a fair chance. This goes for sea, air, rail and road. Tough competition will promote the best over all development. Each of the transport modes have their particular advantages and disadvantages.

Protection of the environment should be brought about by the setting of general standards, e.g. for exhaust emissions, noise, etc., not by detailed and restrictive regulations. There is no doubt in my mind that the automotive industry, as well as other branches, are willing to allocate the necessary resources to find appropriate and adequate technical solutions to any environmental problems. This field contains, in fact, some of the most interesting challenges to industry's research and development departments in the future.

The third factor to influence future developments in the field of goods transport is new technological developments. Probably the most interesting field is that of high-speed trains, where speeds of 100 mph (160 km/h) are possible already today. The technology developed for passenger trains must be transferred to goods trains. This concerns braking systems, bogies and noise suppression.

We also need to develop and apply computer technology to goods transport to a much larger extent. It is possible to eliminate all delays due to customs clearance and other formalities by separating these functions entirely from the physical transport and carrying them out by computer technology. This can be done only to the extent that these often superfluous functions are not themselves eliminated!

2.3 INTERNAL FACTORS

On top of these three sets of more or less external factors come those factors determining future developments in goods transport that industry itself finds necessary in order to meet future requirements. This brings me, of course, to the Scandinavian Link Project. Its basic aim has the objective of radically improving transport links between Scandinavia and the European Continent. Scandinavia is geographically on the periphery of Europe. We must take steps to bring ourselves closer to the Continent.

The project foresees road infrastructure in motorway standard between our major cities and the Continent. This includes, of course, construction of a fixed link across the Öresund between Malmo and Copenhagen. In Denmark, construction of a fixed link across the Great Belt between Zeeland and Funen has commenced in the autumn of 1988. The project also foresees vast improvements in rail infrastructure. All major lines must be upgraded to double lane standard to afford greater capacity and higher speeds. The most important lines are the one from Oslo to the south along the west coast of Sweden to Malmo and on to the Copenhagen airport at Kastrup and Stockholm-Malmo.

Industry requires this dual system, accommodating both road and rail, in order to assure continued free and vigorous competition between the various transport modes. The Scandinavian Link project does not favour either road or rail at the expense of the other.

The results of our studies into the consequences of such a complex of projects show a very satisfactory annual rate of return of about 10 percent. In addition, the competitiveness of business is improved resulting in increased exports.

Even without Scandinavian Link, goods volumes will grow by 40-50 percent by the end of the century. Scandinavian Link makes it possible to handle this increase in traffic with sufficient capacity while at the same time improving the environment. A system of combined transport (road/rail) utilizing new technology for higher speeds and improved infrastructure will be competitive at longer distances and much superior at very long distances. The bridge across the Öresund at Kastrup will ease access to the Copenhagen Airport, thus also improving the competitiveness of air transport. Finally, improved road safety due to motorways replacing ordinary roads will save human lives. Motorways are about twice as safe as other roads.

When the complex of projects entitled Scandinavian Link are completed, the result will be reduced transport time for persons and goods, lower overall logistics costs, an improved environment and safer roads. Improved goods transport is in short one of the very important means to assure continued growth and is thereby to the good of all mankind.

Chapter 3

THE FUTURE OF THE AUTOMOBILE

George C. Eads
General Motors Corporation, U.S.A.

3.1 INTRODUCTION

I am honored to have been asked to participate in this conference on Transportation for the Future, especially during this year marking the 350th anniversary of the first Swedish settlement in America. Recently I was able to attend a conference in Detroit commemorating this anniversary. This conference was titled "Sweden Works: Industry in Transition". The purpose of the SwedenWorks conference was to describe lessons that Swedish industry (and especially the Swedish automotive industry) can teach U.S. industry (especially the U.S. automotive industry). Major presentations were made by members of top management of Volvo, Saab and the new Volvo-GMC heavy truck joint venture. These presentations made a convincing case that General Motors -- and, indeed, the rest of U.S. industry -- can learn a great deal from Sweden's accomplishments.

One statement made during the conference by Thage Berggren, President of the new Volvo-GM joint venture, provides a bridge to the subject of my presentation today. Noting the anniversary being commemorated by that conference, Mr. Berggren confidently predicted that 350 years from now - that is, in the year 2338 - trucks would still be the principal way of moving goods and Volvo-GM would still be a major force in the production of trucks. Talk about taking the long view!!

I am not going to claim today to be able to see 350 years into the future. Instead, I shall be looking no further than the next 25 to 40 years - that is, to the period between 2013 and 2038. To put that sort of forecast period into perspective, it is equivalent to someone looking at today, 1988, from the vantage point of either 1938 or 1963.

Traditionally, the automobile's great strength has been its ability to provide flexible, convenient and comfortable transportation at low private cost. But in the last quarter century or so, other requirements have been added: energy efficiency, reduced emissions, improved safety, and reduced congestion. The challenge today and in the future is to meet these newer needs while maintaining or further improving the former.

Technology will be a critical element in permitting the improvements I will be discussing. But, as Bob Eaton, head of GM's Technical Staff Group, recently observed in a speech to the Government and Industry Conference of the Society of Automotive Engineers, "... there is no room in today's automotive market for technology for its own sake. In a market like today's ... the customer will buy a product only if he feels the value and quality are in line with the price. Any application of new technology must offer the customer more value than the price he has to pay."

To prevent misunderstanding, before going further let me clarify what I mean by the word "auto". Unless I specifically indicate otherwise, you should understand my use of that word to include not only conventional passenger cars but also light-duty trucks and passenger vans. In our own thinking at General Motors, we increasingly view all such personal use transportation vehicles as a continuum, though there are, of course, certain important differences between passenger car and light truck products that justify different regulatory treatment in some circumstances.

As I have already said, it has been the automobile's superior ability to meet the "traditional" needs of a transportation system that has enabled it to become the predominant passenger transportation vehicle throughout much of the world. And as countries around the world have developed, most have strived to increase their "automobility" (to borrow a word from Professor Webber of the Institute of Urban and Regional Development at the University of California at Berkeley; see Webber, 1986.)

But the auto's very popularity in both the developed and developing world has been the source of the objections that are now being raised against further increases in "automobility". For example, in recent years, we have spent large sums on various forms of mass transit in the United States. We have built totally new subway systems in several cities -- San Francisco and Washington come immediately to mind, but there have been others. We have spent billions of dollars remodelling existing subway and light rail systems in cities like New York and Boston. We have also subsidized the upgrading of bus systems in virtually every large and medium-sized American city. We did this knowing fully well that these systems were less flexible and, in many cases, would be less convenient and less comfortable than the private automobile. But we believed that the investment was justified by the potential of these systems to increase energy efficiency, reduce pollution, and reduce congestion.

What has been the American public's response so far? During 1983, the most recent year for which I could obtain data, private autos, station wagons, vans, and pickups still accounted for 82 percent of all the person-trips taken in the U.S.

(MVMA, 1987, p. 51). More significantly, during this same year, eighty-six percent of all workers traveled to work by passenger car, truck, van or other private vehicle. This compares to 5 percent for all forms of public transportation. Indeed, the public transportation share of workers carried to and from work was down from 7 percent in 1969.

The mode whose use had increased most was what the Motor Vehicle Manufacturing Association calls "trucks, vans and other private vehicles". Its share was up from 6 percent in 1969 to 16 percent in 1983. (MVMA, 1987, p. 52.) This increase reflects in part the increased popularity of light truck and van products. Some people who commuted to work alone by car in 1969 were commuting alone by truck or van in 1983. But it also represents the increased use of privately-organized car and van pools -- a subject to which I will return later. I don't want to take anything away from the subway, light rail and bus systems that literally make life possible in some cities around the world. I shudder to think of New York, Boston, or London without their buses and subways. But the important lesson of this recent American experience is that public transit supply does not create its own demand. Unless a transportation system meets important needs better than competitor systems, it will not be used. And if it is not used, its potential for improved energy efficiency, reduced pollution, reduced congestion and improved safety will go unrealized.

In most cases, traditional forms of public transportation just cannot offer the flexibility of the private automobile and light truck or van. As GM's President, Bob Stempel, recently noted in an address to the Society of Automotive Engineers: "American commuters need more flexibility than mass transit can offer. A typical day for the commuter can include dropping the kids off at the day-care center, going to work, picking up the kids, and stopping at the shopping mall or grocery store. That kind of strenuous schedule requires flexibility -- the kind of flexibility that only a private automobile (and, I am sure Bob would agree, light truck or van) can offer."

I am aware that, over a long period of time, the availability of transportation options can change location patterns and that cities like New York and Boston bear the imprint of transportation decisions made in some cases more than a hundred years ago. In some cases, it certainly makes sense to design public transit systems with a view toward effecting such long-term changes. But that doesn't change the fact that, if we hope either today or at any time in the near future to improve the energy efficiency, pollution characteristics, or congestion-generating potential of urban and suburban passenger transportation systems, it is the private auto that will have to form the cornerstone of our strategy.

How has the automobile been doing in responding to these needs?

3.2 ENERGY EFFICIENCY

First let's consider energy efficiency. When one talks about energy efficiency (at least in the U.S.), it is surprising how many people still think of the private car or light truck in terms of the 8- to 10-mile per gallon vehicles of the late 1950's and early 1960's. But except among collectors, these are as extinct as the dinosaurs. Between 1975 and 1985, the test weight of the average car sold in the United States decreased by 24 percent (Kamal, 1987, p. 4.). Powertrains have also improved. Today, we at GM are achieving horsepower and torque levels with four-cylinder engines that were in the past achievable only with our V-6's. Our transmissions are making this power available to the driving wheels with greater and greater efficiency. Tyre friction losses have been reduced, and aerodynamic efficiency increased. As a result, the typical newly-purchased "full size" sedan of today achieved a level of fuel economy far in excess of the "economy car" of twenty years ago. Today, the *average* new GM passenger car gets about 28 miles to the gallon; the average light truck or van, about 22 miles per gallon. And by the year 2000, without any increase in average new-vehicle fuel mileage, the continuing replacement of the oldest, lowest-mileage cars will raise the average of all cars in use in the U.S. by about 35 percent from its present level.

This has been accomplished without severely limiting the public's choice of vehicles. Though passenger and cargo-carrying capacity of the average passenger car purchased today is less than it was a couple of decades ago, the range of capacities (and range of energy efficiencies) available in vehicles offered for sale today is quite remarkable. GM's largest car today, the six passenger, V-8 powered Cadillac Fleetwood Brougham, is EPA-rated at 23 miles per gallon -- one mile per gallon better, by the way, than Volvo's 4-cylinder 240. GM's highest mileage car -- and the highest mileage car sold in America -- is the Chevrolet Sprint which is capable of seating four and gets 56 miles per gallon. (The Sprint is now built in Japan by our Suzuki affiliate, but it soon will be built in North America by a GM-Suzuki joint venture.)

Some in the U.S. have argued that GM should quit (or be prohibited from) producing and selling full-sized passenger cars and instead sell only Sprints or their light truck product equivalents. But would such a strategy make sense from an energy efficiency point of view? I think not. While underlying vehicle fuel economy characteristics are certainly important to attaining high levels of energy efficiency, matching vehicle size to the demand can be even more important. The extremely impressive seat-mile per gallon (or gallon-equivalent) characteristics of large buses, subways, or commuter trains are relevant only if the vehicles are filled or close to it. And even if they are at times, if passengers must travel significant distances in order to access these vehicles, the door-to-door efficiency of these forms of public transit may be much less. Indeed, next to the improvement in average vehicle efficiency, the most important auto-related energy-saving measure that has occurred in the United States in recent years is likely to have been the growth of privately-organized car and van pools.

Having a range of vehicle sizes to draw from is important here also. Based upon its EPA rating, a Chevrolet Sprint when comfortably occupied by two people, generates 112 passenger-miles per gallon. While that is pretty impressive, a GM compact van, which can comfortably seat eight, generates 144 passenger-miles per gallon, even taking into account the reduced mileage that one can expect with a heavier load. And a full-sized GM van configured with a third seat can generate 192 passenger-miles per gallon under the same ground rules - 70 percent more than the Sprint. To be sure, a fully-loaded city bus or subway train will do even better than this. But the chances of finding 2, 8, or even 12 people who live reasonably close to one another and who work at the same place is much higher than of finding the 50 or 60 required to fill up a full-sized bus, not to speak of the much larger number required to enable a subway to operate at reasonably close to full capacity.

Little wonder that in the article I have already referred to, Professor Webber suggests that if we could increase commuter auto occupancy from its current 1.4 passengers per vehicle to only about 1.6 passengers per vehicle, we could achieve operating energy efficiencies equal to those actually being achieved today by modern electrified railroads or subways. This would be even higher than 30 to 60 passenger buses, whose average passenger load today is only about ten passengers. It would also substantially reduce pollution and congestion.

This is one of the major reasons why GM has opposed, and continues to oppose, the U.S. Corporate Average Fuel Economy (or CAFE) regulations as well as the so-called "gas guzzler" tax. The former require manufacturers such as GM to achieve a certain average number of miles per gallon for all new vehicles they sell, regardless of vehicle size. The latter are designed to make vehicles having fuel economies below a certain level prohibitively expensive to purchase. (Actually, the CAFE regulations are more stringent than I have indicated. Since, in the name of "protecting" domestic jobs, they require us to average separately our North American and overseas-produced vehicles, we cannot now count the 56 mpg Sprints we sell toward meeting CAFE.)

There are, of course, improved technologies that could improve fuel economy further. As I will note shortly, we are working to develop and implement them. But for the immediate future, the simplest (and, increasingly, the most likely) way that we would meet tighter and tighter CAFE requirements would be to restrict the mix of cars and light trucks we offer. This not only would cause severe economic disruption. It also might well reduce the actual level of fuel economy that the country is able to achieve by reducing the useage of car or van pooling or by encouraging people to hold onto their less efficient older vehicles longer.

Let me mention briefly the new technologies we are working on at GM to improve the energy efficiency of our vehicles. To reduce structural mass without reducing vehicle capacity, we are exploring the use of new materials. Engineers at GM research have developed and built an advanced space frame (that is, the skeletal structure of the vehicle on which external panels are mounted) built entirely of composites. This space frame is 20 percent lighter than its steel counterpart and meets all production and crashworthiness criteria. We are also working to apply ceramic components to automobile engines. For example, ceramic components in a gas turbine could improve efficiency, permitting the size and weight of the engine to be reduced by as much as fifty percent without any loss of power. Finally, we are working to improve transmission efficiency -- the avenue that many of our engineers believe holds the greatest near-term potential for energy efficiency improvement (Kamal, p. 17).

However, new car fuel costs per mile, adjusted for inflation, are already at an all-time low in the United States. Therefore, it is likely to be hard to convince the American motoring public to pay the extra cost that would be required to put vehicles embodying such advances into production in the near future. Government "incentives" could be used to make current cars so expensive that the new technology might be relatively more attractive. But increasing the cost of cars does not make people more inclined to buy them. What about increasing the price of gasoline through taxes? This certainly would increase the value of fuel efficiency to the consumer. But would it make sense? Although fossil fuels can be expected to increase in price over the longer term, it is not geology, but politics, that is the source of today's principal energy worries -- namely, the sort of "oil shocks" that we experienced in the early 1970's and again at the end of that decade. But if it is supply disruptions you are concerned about, the arguments in favor of retaining flexibility in the privately-owned vehicle fleet are substantial. In the two oil shocks, the unused seating capacity of larger cars represented an important buffer that absorbed surges in demand at minimal inconvenience.

While I am mentioning energy availability, I should point out that in addition to working on ways to improve vehicle fuel economy, we are also continuing to explore the application of alternative fuels. We recently donated to the California Energy Board a Chevrolet Corsica that can use any mixture of methanol and gasoline from 100 percent of the former to 100 percent of the latter. And we have also put methanol-powered buses into normal operation in several cities across the United States. We expect to have more than 60 of these demonstration buses in operation by the year 1990.

But it is important to keep this work in perspective. In the near term, we view methanol not as a general substitute for gasoline but as a potential gasoline supplement in those areas of the country suffering from otherwise intractable air pollution problems. As far as the U.S. is concerned, the economics of methanol as a general gasoline substitute is quite unconvincing right now. However, we do see more favorable economics for other parts of the world with a surplus of natural gas, and potentially also for the U.S. if natural gas proves to be sufficiently plentiful.

To sum up on energy, while it is important to continue our search for economical alternative fuels and for technologies that can help us improve fuel efficiency, we should not overlook using economic incentives to raise the intensity of vehicle use, especially for commuting. Restricted High Occupancy Vehicle (or HOV) lanes, freedom from bridge and road tolls, and special parking incentives deserve greater consideration, certainly in contrast to more stringent CAFE regulations, higher "gas guzzler" taxes, or the massive subsidization of additional "heavy rail" transit systems.

3.3 POLLUTION EFFECTS

Let me turn now to the second of the major new objectives I cited earlier, the reduction of pollution. There is no denying that motor vehicles have been (and in some cases still are) the largest single source of air pollution in many places throughout the world. There is room for dispute about the ambient levels at which the health effects of certain motor vehicle-related pollutants begin and about the precise nature of the linkage between motor vehicle emissions and ambient pollutant levels. But the time has long since passed when anyone could seriously deny the need to control motor vehicle emissions. The relevant issue today is not "whether" but "how much".

As with fuel economy, many people are unaware of the progress that has been made in reducing new vehicle emissions. Likewise, they are unaware of the degree of further progress that is assured merely from diffusion throughout the vehicle population of the technologies that already are in common use on new vehicles currently being produced in the United States. As compared to uncontrolled vehicles, new passenger cars sold today in the United States emit 96 percent less carbon monoxide, 96 percent less hydrocarbon, and 76 percent less nitrogen oxide, not merely when new, but after 50,000 miles of use. And in spite of projected increases in the size of the fleet and vehicle miles traveled, the effects of fleet turnover are expected to continue reducing vehicle emissions for years to come even with no change in current standards. For example, total auto-related hydrocarbon emissions in the U.S. are expected to continue declining through 1999, eventually bottoming-out at a level between 26 percent and 52 percent lower than today.

But despite these improvements, many urban areas in the U.S. today still fail to meet federal ambient air quality standards for certain auto-related pollutants. It is here that one unavoidably gets into questions of what "failure" to meet standards really means and how significant it is as a matter of public health. I shall be happy to talk about these matters later with anyone who wishes. But at this point just let me assert that, aside from a few areas of the country (one of which is the Los Angeles basin), the principal auto-related air pollution "problem" is the slowness of fleet turnover, not the emission characteristics of new vehicles. And, as I have already indicated, the best approach to dealing with these "special case" areas is not to tighten (either directly or indirectly) the emissions standards applying throughout the country.

What does technology have in store for us regarding improvements in emissions? The technologies that I have already mentioned with regard to improving fuel economy are also emissions-reducing technologies, since reduced fuel use generally translates into reduced emissions. But beyond this, the technology cupboard is not very full. Two remarkable pieces of technology -- the catalytic converter and the micro-processor adapted to engine controls -- have been responsible for the bulk of the progress that has been made to date in reducing auto emissions. And while our engineers continue to look for other technologies having similar emissions-reduction potential, little has appeared on the immediate horizon. Rather, they plead for a period of stability in the regulations so that our ability to improve the in-use performance of these technologies can be enhanced. Such improvements only come when we have such periods of stability and can concentrate on "fine tuning" rather than revolutionary change. That is the reason why GM is opposing further tightening of the Clean Air Amendments.

3.4 SAFETY

Now let me turn briefly to safety. In recent years, the rate of fatalities per vehicle mile has continued the general downward trend it has displayed almost as far back as records have been kept. (MWMA, 1987, p. 90.) And, although the effectiveness of certain individual federal auto safety standards continue to be a matter of legitimate dispute, recent research has largely settled in the affirmative the question of whether federal motor vehicle safety efforts have contributed to this improvement. (Crandall, 1986, Chapter 4; Graham, 1984.)

What lies ahead? One of the most important advances in vehicle hardware is the rapid incorporation of anti-lock brakes as standard equipment on cars and light trucks. I suspect that this may be occurring more rapidly than many might think.

On the issue of occupant restraint systems, though the debate has long been settled elsewhere (generally through the passage of mandatory seat belt laws), the U.S. still continues to struggle. Our engineers remain convinced that the three-point shoulder harness and seat belt, properly worn, provides vehicle occupants the best protection in the widest possible range of crash circumstances. The problem, of course, has been to get vehicle occupants in the U.S. to use these systems. We at GM have favored seat belt laws such as exist in many countries throughout the world. However, after a nearly 25-year struggle, the Federal government purports to have made a final determination to require "passive" restraints. We and most others who sell vehicles in the United States plan to meet this requirement through the installation of *supplemental* air cushion restraint systems in our larger cars and "passive belts" in our smaller cars (I stress the word "supplemental" because these systems are intended to be used *in connection with* a seat belt). We hope that these supplemental air cushion restraint systems prove in practice to be as effective and as acceptable to vehicle occupants as the government expects.

In the future, such things as collision avoidance radar and "heads up" display systems may provide additional motor vehicle safety opportunities. This model year we will be the first manufacturer to introduce "heads up" displays into production vehicles, though in extremely limited numbers. As with all such technologies, the basic question still to be answered (beyond that of whether the technology is really effective in practice) is whether it can be introduced into production vehicles at a cost that is attractive to consumers.

3.5 CONGESTION

Finally, let me turn to the issue of congestion. There is a growing appreciation around the world of the need for automakers to devote increasing attention to the problem of congestion. It is becoming clear that merely building more roads is not the answer to our growing congestion problems -ways must be found to utilize our existing road systems more efficiently.

There are a number of competing approaches. In several countries, groups are working on a concept that might be called the "fully-automated highway system". In this system, traffic flow and speed would be determined and controlled by monitors and power sources in the highway itself. But such systems would require an astronomical investment and appear to be at least several decades away.

At GM, we think a more promising and more feasible approach is to establish a direct two-way communications link between urban traffic control centers and the vehicle itself. This would instantaneously provide the driver with up-to-the-minute information on traffic and roadway conditions and alternate routes and, eventually, a wide assortment of information services. We are participating with the Federal Highway Administration and the California Department of Transportation (Caltrans) in a three-year research project called "Project Pathfinder".

The goal of this project is to gain empirical data and real-world experience on the effectiveness of electronically linking all available traffic condition information with a computerized mapping display located on the vehicle's dashboard. The system will be tested in the heavily-travelled Los Angeles-Santa Monica corridor. GM will provide 25 Oldsmobile Delta 88's which will be fitted with these electronic maps to receive and display traffic and accident information fed directly from Caltrans' operations center. The electronic map screen will not only provide up-to-the-minute information, but will also display routing alternatives to avoid congested areas. This will be the first time such a system has actually been tested in the United States.

We are aware that Europe and Japan are going much further and much faster in research and testing in this area. Some of the later chapters in this volume report on these developments (see Chapters 8 and 9), and I look forward to reading them. An important characteristic of these efforts is that they involve consortia of companies working together with government.

Thus far, the U.S. auto companies with operations in Europe have been excluded from Europe's electronic traffic information project known as Prometheus. For this reason, and also to accelerate work on this topic in the United States, Bob Eaton proposed in his speech to the SAE Government and Industry Conference that the U.S. government and American automakers explore jointly undertaking work to assure that the U.S. is not disadvantaged in this race. He also suggested that the 1991 expiration of the basic legislation funding the federal highway system to be seen as an important opportunity to guarantee the availability of monies to assure that this country's highway system is brought up to par with those being developed today in Europe and Japan.

Let me close by mentioning briefly another challenge that is being created by all this new technology I have been describing - the challenge of effectively integrating all the various complex systems that are either already installed or are now being developed for the automobile of the future. Looking just a few years ahead, it is conceivable to picture quite a few (if not all) automobiles containing not only the electronic emission and engine controls that are rapidly becoming standard in the U.S. today, but also anti-lock brakes, active suspension systems, traction control systems, collision avoidance systems, on-board diagnostic systems, heads-up displays, and electronic navigation and information systems (By the way, given this, it is little wonder that we see the automotive electronics market growing from $15 billion today to $35 billion by the mid-1990s, and the electronic content of the typical vehicle growing from an average of $500 today to more than $1200, by 1995). But for this to occur, it will be necessary to integrate many of these independent systems.

An example concerns the benefits that can be obtained from the use of multiplexing. We have shown that, through multiplexing, it is possible to reduce the number of wires in one of our car doors from 54 to 4. Multiplexing saves weight, space, assembly time and costs as well as improving reliability and ease of service. But to take full advantage of multiplexing, a systems engineering approach to the automobile's electrical components is essential.

One of the principal reasons that GM acquired Hughes a couple of years ago was to be able to tap its wealth of systems engineering experience. Presently, Hughes is working closely with our research and engineering staffs, as well as with our car groups, to both train our people in systems engineering and to integrate systems engineering methodologies into our product program management process. We believe that this will not only achieve a considerable reduction in concept-to-production time but also enable us to produce vehicles which better satisfy the needs of the customer.

3.6 WHAT OF THE FUTURE

In just a few short years, the automobile has been transformed from a relatively simple, largely mechanical product to a product embodying the latest microelectronic technologies. This transformation process is continuing. Manufacturers around the world, working both on their own and with the help of their governments, are vying for leadership; realizing that such leadership represents the key to producing products that not only satisfy the traditional needs of flexibility, convenience and low private cost, but also the newer needs of energy efficiency, low emissions, enhanced safety, and reasonable freedom from congestion. While I cannot predict what will be happening 350 years from now, I can say that the competitive struggle I have just mentioned will help to assure the predominance of the automobile as a mode of passenger transportation for at least the next 25 to 50 years, if not well beyond.

REFERENCES

Robert W. Crandall, Howard K. Gruenspecht, Theodore E. Keeler and Lester B. Lave, *Regulating the Automobile* (Washington: The Brookings Institution, 1986).

John D. Graham and Steven Garber, "Evaluating the Effects of Automobile Safety Regulation", *Journal of Policy Analysis and Management*, Winter 1984, pp. 206-224.

M. M. Kamal, "Future Directions in Automotive Bodies and Powertrains", Research Publication GMR-5697, General Motors Research Laboratories, March 5, 1987.

Motor Vehicle Manufacturing Association, *MVMA Motor Vehicle Facts & Figures, '87* (Detroit, Michigan: 1987.)

Melvin M. Webber, "Automobility for Everyone", *Urban Resources*. Vol 4, No. 1, Fall 1986.

Chapter 4

THE EMERGING C-SOCIETY

Åke E. Andersson and Ulf Strömquist
Institute for Futures Studies and TEMAPLAN AB, Sweden

4.1 EUROPE AND THE FOUR LOGISTICAL REVOLUTIONS

Industrial and regional development in Europe has been shaped by the growth and change of logistical systems, i.e. the slow but steady evolution of logistical infrastructure. Logistical infrastructure is in this context defined to be equipment and networks used for the transportation and distribution of commodities, people, information, and knowledge. The logistical system of an economy also includes the infrastructure within urban nodes.

Shortly after 1.000 A.D., a slow but steady improvement of the transportation network started in the Mediterranean area, followed thereafter by a similar development in Northern Europe. The improvement of sailing technology and the successive opening of harbours and rivers for trade was completed by the opening up of the Champagne fairs, which were controlled by Italian tradesmen. The southern Mediterranean trading system was integrated with the northern trading system, in particular the Hanseatic league, for the first time. Hence, the first European economic system was formed. It implied the rise to world economic dominance of a number of coastal regions of Europe: Bergen, Lubeck, Bruges, Toscana and Venice. On the basis of a new logistical network, a new integrated western European economic network was formed.

Near the beginning of the 16th century this first logistical era was drawing to a conclusion. But a slow and steady development of the transportation technology, and the introduction of new banking techniques, paved the way for a new and more extended economic system. The introduction of the Caravelle ship made it possible to extend international transportation and communication beyond the limits of the earlier economic system. Lisbon replaced the Champagne fairs as a connection for long distance trade and transportation. It became an ideal common connection point in the new network for transportation between different parts of Europe and the Americas. But Lisbon was a very unstable connection in the new international economic order. It was soon to be replaced by Antwerp of the Flanders. With the development of the new banking system and the innovations in banking developed by Amsterdam, the final stable focal point was shifted once more to the city of Amsterdam - the new capital of the world economy after the second logistical revolution.

The first logistical revolution was primarily caused by improvements to coastal transportation systems. The second logistical revolution included the introduction of more complex systems for interaction. It was caused by improved possibilities of ocean transportation and by dramatically improved systems for transactions involving credits designed to serve long distance and long term trading expeditions.

The third logistical revolution - the industrial revolution - was not caused by the introduction of machinery, as it has often been assumed. It was rather the consequence of a slow and steady change of transportation and transaction possibilities at the intercontinental and intracontinental level. Great Britain had developed a qualitatively superior transaction system based on the formation of the central Bank of England. Furthermore, Great Britain had secured successively safer and faster possibilities of North Atlantic transportation and trade. It became more and more obvious that the future of the European economy would be based on the exploitation of the European integration with North America rather than on trade with the East Indies.

The new transportation and transaction network was to be based on an integration of Western Europe and the eastern parts of North America. Southern regions were thus naturally favoured more than others by this new logistical era. Among the most favoured regions were the new British centers of the new industrial system. This set included Manchester and Liverpool. These became more closely connected by the new canal and railroad networks. All this made it possible to utilize machinery equipment in plants and production units which exploited economies of scale. Hence, the industrial revolution was a subtle combination of intraplant and intercontinental logistical systems.

It is obvious that the three major eras in European economic development over the last 1800 years have been influenced by slow and steady changes to the transportation and communication systems. In all these cases the responses from the economic system have been multi-dimensional:

A. an almost explosive expansion of trade and transportation,
B. rapid expansion of income and wealth in specific regions and occupations,
C. immigration to these sets of occupations and regions,
D. new social and political tensions,
E. formation of new political power groups, and
F. outbursts of creativity in economic life, science, and the arts.

4.2 THE TRANSFORMATION OF THE LOGISTICAL SYSTEMS

The third logistical (or industrial) revolution was based on increasing possibilities of sea, canal and railroad transportation, i.e. steam technologies on land and waterways. The basis of the new industrial system was an infrastructure for transportation and communication based on a few large links connecting a relatively small number of major harbours and switchpoints in the railroad systems. This was a transportation network designed to favour trade in bulk commodities. Although the new economic system attracted new production technology, it was a production system based on the exploitation of material resources, energy and massive inputs of unskilled labour. This economy made prudent use of the scarce transportation, communication and knowledge resources. The hierarchical, large scale structure of the transportation, communication, and knowledge infrastructure required similarly hierarchical and large scale production systems (which seen from a theoretical perspective are local logistical systems).

The relative role of water and railroad transportation networks is, however, slowly but steadily decreasing during the 20th century. This can be shown by the diagram on the next page indicating the relative share of total network links of different transportation systems from the beginning of the 1850s until 1985 (with a short-term projection into the future). Table 4.1 records recent trends in Sweden.

Table 4.1 The value of Swedish import flows distributed over transportation modes

	1960	1980	2000
Sea and rail systems	82	42	15
Road and air systems	18	58	85

Source: Andersson, Strömquist (1988)

The following diagram, which is computed for the U.S.S.R., is highly indicative of the development throughout Europe. It is the same pattern as we have witnessed in the USA (see Chapter 5), but with a time delay of one to two decades.

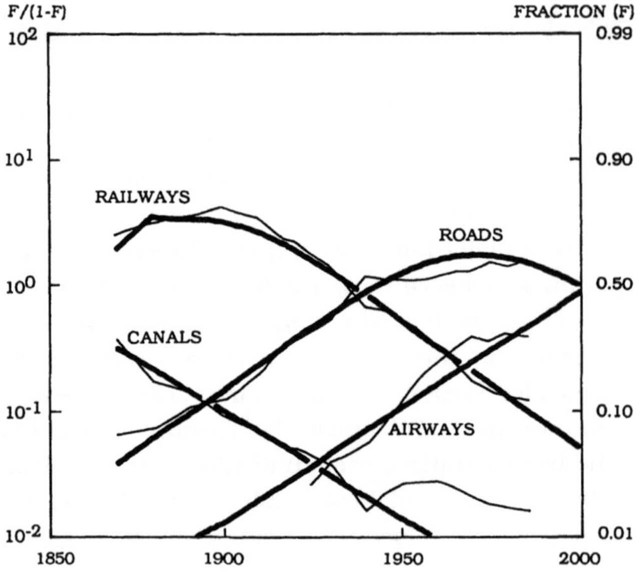

Figure 4.1 Development of transportation infrastructure in the USSR by shares in total length.
Source: Grübler (1988)

The communication systems are following a similar pattern. Data from Sweden are quite revealing in this respect. Sweden has experienced the fastest expansion of telecommunications in Europe (see also Chapter 6), a fact primarily reflecting the institutional organization, permitting competition between postal communication and the telecommunication system. The two diagrams in Figure 4.2 show that the expansion of truck transportation develops in parallel with the growth of telephone calls.

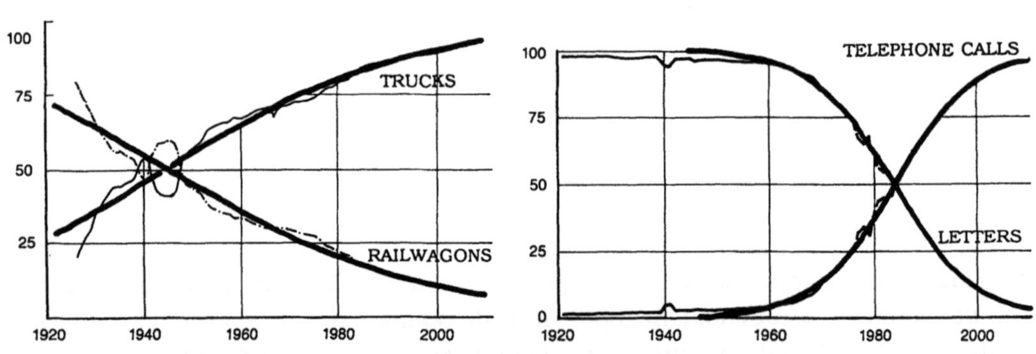

Figure 4.2 Letters and railwagons replaced by telephone calls and trucks. Sweden
Source: Andersson & Strömquist (1988)

A steady expansion of the educational system and complementary "on the job training" has provided another foundation for a new production and distribution system in Europe. Knowledge handling is becoming a more important economic activity. Its steadily increasing share is associated with the complete renewal of the economic system towards dynamic product competition.

A new economic system in Europe will thus be based on a new infrastructure consisting of:

* a road network, primarily consisting of highways,
* road network equipment in the form of adaptive and fast trucks and cars for transportation of high value commodities and people, computerized traffic, freight and travel guidance and control systems,
* advanced and efficient telecommunication networks with high fixed costs for installations and low variable costs with respect to both volumes of information and distances of transmission,
* information technology equipment which can be used interactively on the new telecommunication network. Among such equipment we will not only find large and small computers, but also computer-based facsimile equipment as well as mobile telephones, videotex, etc,
* an extensive and much more competitive air transportation system consisting of airports, traffic corridors, computerized traffic guidance systems, etc. Important users of these systems are knowledge handlers, administrators, and other advanced personnel. They all constitute contact-dependent users of air transportation.

This communication and transportation infrastructure provides the arena onto which the new economic system of Europe is emerging. Such a new economic order will affect Europe in a revolutionary way, following in the footsteps of earlier logistical revolutions.

4.3 FROM HIERARCHICAL TO SELF-ORGANIZING NETWORKS OF PRODUCTION SYSTEMS

The logistical systems of the industrial era were based on large and homogenous flows between ocean ports and large scale manufacturing systems. These were well connected with each other by means of the ports and interconnecting railroad transportation networks. Economies of scale in the transportation networks required economies of scale and a hierarchical production and distribution system in order to be efficient.

The new interactive networks for the transportation of goods, services, and knowledge, and the communication systems for transmission of information and routinized messages, are much more dense, admit much more frequent and small scale deliveries and are much more adaptive in time and space. This implies that there is no longer any strong requirement for the classical hierarchical production and distribution system. Neither is there any strong requirement to use hierarchical systems for control and administration of these systems. The new industrial paradigm of organisation can be called "Self organized networks of R&D, production, and distribution units of corporations". Digital Equipment Corporation is a good example. According to the new guidelines for this corporation, individual units are constrained to be smaller than 400 employees, and these individual units are free to self-organize within small networks, as long as they fulfill corporate profitibility, liquidity, and marketing guidelines.

4.4 FROM LOCAL TO CONTINENTAL AND GLOBAL RESOURCE SUPPLY

The classical industrial system was and is built on location patterns which are sensitive to the availibility and procurement possibilities of local and regional resources. An example is the furniture industry, which was traditionally based upon the availability of wood, textiles and other regional raw materials, as well as labour and regionally accessible semi-finished inputs.

This is now rapidly changing. Firms like IKEA are introducing Just-In-Time and other elements of systems organization to their production. With the use of the new transportation and communication technologies, production of components and semi-finished inputs can be allocated to the most advantageous parts of the European continent. Only resources for development and design activities and marketing are kept at the center of the organization. This means that corporations like IKEA and Digital are changing to a network of relatively small production and distributing units, where each unit is located with respect to the availability of either continental resources or market potentials.

Again, we find an increase in the size of the corporate structure combined with decreasing size of individual production or distribution units.

4.5 RISE OF C-REGIONS, DECLINE OF M-REGIONS

During the industrial era in the United States and Western Europe, there was a steady decline in regional inequality of income per capita. Studies by Williamson, Easterlin and Andersson have all shown that in the advanced stages of industrial civilization, there is a steady decline of all measures of regional inequality. In practically all observed developed countries there has been a convergence of regional percentage shares of population towards regional shares of national income. Such a development should not be expected to continue during the decades to come.

All the three former logistical revolutions have been acompanied by fast growth of new centers of regional and industrial expansion. It is to be expected that the emerging fourth logistical revolution will cause new regional tensions. In the United States one can observe the emergence of new C-regions, where the expansion has been motivated by the abundance of C-resources in these regions. The Route 128 region outside of Boston is an early example of such a corridor region expanding in terms of employment, income and quality of life at a much faster rate than other regions which are less developed in terms of communication capacity, cognitive skill, knowledge availability, or supply of creative and cultural capacities. Other examples of American C-regions are the San Francisco bay area (with Silicon Valley), and the San Diego-Los Angeles development corridor.

In Europe, the most prominent examples of similar changes in the regional development pattern are to be found in the Cambridge-London corridor, the Uppsala-Stockholm corridor, and the Munich, Stuttgart and Bonn-Düsseldorf regions of Germany.

In Sweden most of the growth in the period 1975-1988 has been channeled into the Uppsala-Stockholm corridor and the C-regions (equipped with old or new universities and air transportation capacity). A recent econometric study (Andersson, Anderstig and Hårsman, 1988) shows that the best explanation of regional productivity and income development is the relative abundance of university, airline and road transportation capacity of the region. According to this study, rapid development of a region is strongly favoured by a combination of university and other research facilities growing synergetically with communication and fast transportation capacity. At the same time, many studies indicate that there are no special advantages in metropolitan regions during the later stages of the industrial era (i.e., the period 1960-1975) among the OECD countries.

Other studies have shown that development paths based on modern infrastructure are not necessarily of a very centralistic nature. It has been shown that smaller universitites (with full Ph. D. programs) are as productive and creative as the larger universities. Some of these smaller R&D units are examples of recent regional development policies. This is an observation further reinforced by the fact that the best American universities are in fact comparatively small. Examples of European policies aiming at the decentralisation of university capacities, combined with highway programs and general improvement of transportation and communication capacity are thus consistent with a modern regional policy.

Similarly, the recent tendency of Nordic regional policies oriented to the expansion of new university capacity in sparsely populated parts of the Nordic countries, and the expansion of air transportation and communication capacity, combined with technology diffusion programs are also completely in line with a rational regional policy for the C-society.

Some problems are now emerging in rural areas located far away from any such modern infrastructure. Innovation diffusion in this new stage of economic development must proceed rapidly, if lagging development should not become a major problem. However, innovation and diffusion of the modern information technologies are not as swift as often assumed in policy making documents. The critical element in the development of rural regions and their urban-like nodes is the effort necessary to link themselves efficiently via networks to the major nodes of the C-society.

In the EEC countries, household expenditure on transport and communication is estimated to lie between ten and seventeen percent of disposable income. This share might reach 20-25 percent after the year 2000. Direct costs for goods transport are typically of the same order of magnitude as energy costs, i.e., about 8 percent of GNP. However, generalized transport costs are over 30 percent in Sweden if the costs of handling, information processing, storing and the capital cost of inventories are included. These generalized transport and distribution costs, often called the costs of logistics, have since the third logistical revolution increased due to increased specialization (more steps in the logistical chain), larger geographical markets of inputs and outputs, and a lower speed of automation and rationalization in the distribution sector compared with the production sectors.

4.6 TRANSFORMATION OF COMMUNICATION TECHNOLOGY AND ITS CONSEQUENCES FOR REGIONAL DEVELOPMENT

Communication technology is now rapidly changing. All over the world computers are being linked with each other by modern telecommunications networks. Fax machines are expanding in most countries at a rate exceeding 35 percent per year. Mobile telephones are growing at comparable rates. The very sparsely populated Nordic countries are the most prominent examples of rapid expansion of this modern technology. In some sparsely populated rural areas, they are cheaper to install and use than ordinary fixed location telephones. For small scale service sector production in rural areas, it is probably the single most important productivity-enhancing technology of the coming decades.

The very fast development of new communication technologies based on telecommunicating, small computers, fax transmission equipment, and mobile telephones has increased the possibilities of relocation to and growth of industrial units and service activities formerly unsuitable for rural regions. The real development problems of sparsely populated regions are related to key shortages in the supply of highly qualified labor and the lack of accessibility to air, road or railroad systems for the transportation of high value commodities and people.

4.7 THE PATTERN OF LOGISTICAL AND INDUSTRIAL REORGANIZATION

The arena of networks for the movements of commodities, people, knowledge and information is slowly changing in the developed world. In any logistical choice situation, there is now a multitude of options available for the allocating and locating firm.

In the period of the industrial era, two types of infrastructure dominated the stage - the railroad network and the system of harbours, steamships and shipping lines. The sea and the rail networks were complementary with common terminals in certain ocean harbours. At these switching points or saddles in the international trading system, there were great advantages of locating large scale, logistically sensitive plants. Because of the indivisibilities and other scale economics of railroads and sea transportation networks, there emerged a necessity for economies of scale in the processing units of the industrial logistical system as well. In summary: if one dominant part of the logistical system is designed for economies of scale, the other systems will tend to become dominated by similar considerations of economies of scale. A harbour designed for bulk transportation needs production systems designed for bulk production.

During the industrial era, routinized mass production was the key to achieving economic efficiency. This, in its turn, implied that informaiton was routinized and the need for information exchange was limited and "one-way". With the need to produce large amounts of standardized commodities there was also a limited requirement for research and development and other forms of knowledge production. Thus the mass production units and the mass transportation networks were driven by very limited amounts of slow information and knowledge exchange. As these information and knowledge flows were concentrated and one-way, there was a definite point of concentration of the generation of information and knowledge. The corporate headquarters tended to be located close to the largest processing plants. Economies of scale were supplemented by hierarchical organization of this decision making.

In the new, dense multi-layer logistical networks with their fast but thin links, there is no economic efficiency of a strict hierarchical structure with large scale processing units. The new logistical systems are being transformed into self-organizing networks of processing, knowlede-enhancing, distributing, and information handling units. It is self-evident that such a logistical system - based on fast and well-coordinated, small scale flows of commodities and people - will be using the adaptive and fast modes of transportation (i.e. cars, trucks, and aeroplane transportation). These are systems with an average load less than one percent of the average transportation load for freight during the industrial era.

With the scarcity of transportation and communication capacity during the industrial stage of development of the world economy, the organization of space was determined by transportation frictions. Spatial interaction tended to be based on contiguity of the interacting units. Terms like "central place" and "intervening opportunities" were of great relevance in the analysis of locational patterns. With the new, multi-layer, dense grids of different networks, contiguity tends to be replaced by *similarity of demands and complementarity of resources*. One example of the new networks is the R&D and production networks of the biotechnical complex. Approximately ten centers of excellence in different parts of the world are interacting intensely with each other with very few multiplier effects or interaction relations on places in their regional vicinities. Within these global or continental networks of firms interacting at increasing distances, there is a very strong tendency towards industrial cooperation and merger. Though there are very limited advantages of large scale production at the processing level, there is an increasing advantage of corporate economies of scale, due to the publicnature of research and development and marketing strategies. Thus there is a simultaneous increase in the average size of corporations and a decrease in the size of the individual production and distribuiton units of each corporation.

The classical spatial organization of firms was also hierarchical. In the case of multinationals, there was a tendency to have a continental level, a national level, a district level, and a local level. The interaction between these different levels was handled by hierarchical information and commodity flows with a system of intermediate storage capacities. The emerging networks have no need of these intermediate levels. Storage systems are being concentrated at the continental level and, in some cases, all relations between producers and users are handled by direct door-to-door transportation systems.

Such a tendency is further reinforced by the transformation from large scale production of standardized commodities to customizing. With the knowledge- and information interaction between producer and user, there cannot be any intermediary storage systems. This emergent C-society calls for fast, direct modes of transportation, individually and in combination. Accessibility to such transportation possibilities will be an essential ingredient of the leading cities in the network era of the near future.

REFERENCES

Anderstig, C. and Mattsson, L-G (1989) *Interregional Allocation Models of Infrastructure Investments.* Office of Regional Planning and Urban Transportation. Stockholm.

Andersson, Å. E., Anderstig, C and Hårsman, B (1988). *Knowledge and Communications Infrastructure and Regional Change.* Office of Regional Planning and Urban Transportation. Stockholm.

Andersson, Å. E. and Strömquist, U. (198). *The Future of the C-Society*, Prisma, Stockholm (in Swedish).

Grübler, A. (1988). *Rise and Fall of Infrastructures, Dynamics of Evolution and Technological Change in Transport.* Ph D Dissertation, Vienna: Technische Universität.

Johansson, B. (1987). *Information Technology and the Viability of Spatial Networks*, Papers of the Regional Sience Association vol 61, pp. 51-64.

Karlsson, C. (1988). *Innovation Adoption and the Product Life Cycle.* Umeå Economic Studies No 185, University of Umeå.

Kearney, A. T. (1986). *European Logistics Productivity Survey*, Brussels.

Swyngedouw, E. A. (1988). *Just-In-Time Manufacturing and Regional Industrial Structures*, WP, IIASA, Laxenburg, Austria.

PART II

THE TRANSPORT-COMMUNICATION NEXUS: SOME EVOLUTIONARY TRENDS

Chapter 5

EXPANDING TERRITORIES: Transport Systems Past and Future

Nebojsa Nakicenovic
International Institute for Applied Systems Analysis, Austria

5.1 INTRODUCTION

Transport and communication systems are the elements which integrate human activities in space and time. They record our past and will be the important determinants of our future. As a rule, those forms of infrastructure which are used for the movement of people, goods and information diffuse slowly and span many decades from their first introduction to obsolescence. Some of them are almost immortal, even though they often provide different services from those originally intended. Obsolete canals were reused to build railways a century ago in England. Ancient Roman roads have often been buried beneath modern highways. Old harbours are being converted into modern commercial and residential areas.

In this sense, many forms of infrastructure often serve as "rights-of-way", as actual transport systems replace each other in the eternal quest to increase the quality of service, reliability and speed, and to make the movement of people, information and tangible goods more efficient and convenient. While old forms of infrastructure are frequently recycled, new systems which provide even greater possibilities are also introduced so that diversification and productivity usually increase in time. Thus, complexity increases, resulting in numerous interlaced and overlapping niches occupied by competing modes of passenger travel, information channels and transport modes for goods and services.

The history and vintage structure of these systems will, to a large extent, determine and constrain the potential for introducing new systems. The development of new systems is, at the same time, a long-term process. The regularities in this evolutionary process that are reflected in the dynamics of change provide insights into possible future developments.

5.2 TRANSPORT INFRASTRUCTURES

One clear trend emerges from historical analysis: transport systems have become increasingly fast and more productive (when measured in terms of ton-kilometers and passenger-kilometers). The first major improvement occurred with the age of canals. They represented a fundamental construction effort toward reducing natural barriers in order to connect coastal and inland waterways into one integrated transport infrastructure. At the same time, canals were a powerful motor for the industrial age. Waterways allowed for new flows of goods, unprecedented exchanges between regions, specialization of labor, and access to more distant energy and raw material resources.

The modern age of canals started about two centuries ago and lasted almost one hundred years. By the 20th century most national canal systems were in place and many links had already been decommissioned. Eventually they yielded to the vicious competition of railroads.

The first railways were constructed in the 1830s and were able to extend the range, speed and productivity of canals. More important, perhaps, was the capability to overcome even more imposing natural barriers. Bridges and tunnels were built for canals, but railways were capable of following traffic and freight demand more directly. Wherever demand existed it was almost always possible to build a railway line. In time, North America and Europe were overlayed with elaborate networks of railway systems. Together with railways a new era of coal, steam, steel and telegraph began. The great railways era lasted until the 1920s.

Around the turn of the century the automobile was born. It became the symbol of modern industrial development along with oil, petrochemicals, electricity, the telephone and (Fordist) manufacturing. Paved roads reduced the time-space dimensions of modern societies. Speed and performance increased once again. This flexible and individual mode of transport became affordable for a wider social strata, and it was not until the last two or three decades that the disadvantages of the automobile have become socially transparent. This illustrates the extent to which the automobile age is perceived as one of the preconditions for modern industrial development. So much so that the car became almost a synonym for individual freedom and mobility.

Figure 5.1 illustrates the development of the four major transport systems in the USA, represented by the growth in length of the respective infrastructures: canals, railways, surfaced roads and airways. The length of all four has increased by more than four orders of magnitude during the last two centuries. Each successive mode of transport expanded into an infrastructure which was ten times larger than the previous one. The first canals were built in the 1780's, reaching a total length of 4,000 miles by the 1870's before saturation and decline; thus the expansion of canals lasted about 90 years. The first railroads were built in the 1830s and

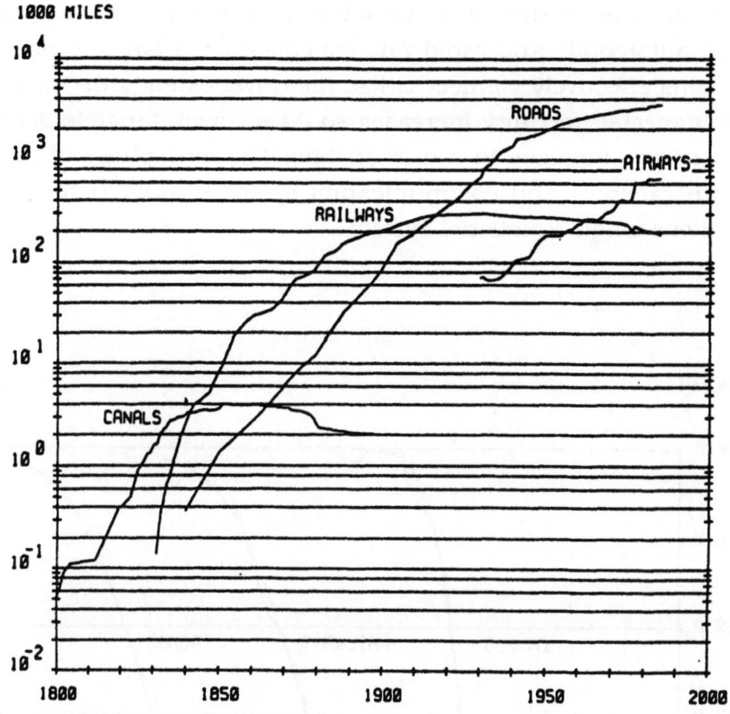

Figure 5.1 Length of Canals, Railroads, Roads and Airways in the USA. Source: Nakicenovic (1988).

saturation started in the 1920s; again about 90 years later. By 1929 the total length was more than 300,000 miles (480,000 km). Thus, the railways saturated at almost ten times the level of canals. Since then rail infrastructure has undergone a phase of rationalization, a loss of market volume both for freight and passenger transport, and a decrease by about one-third to some 200,000 miles (320,000 km).

The first high-quality roads of significant length were introduced a century ago. Today, surfaced roads are approaching saturation with about 3.4 million miles, again more than a factor of ten higher than the maximal length of railways. Each successive transport infrastructure has not only been an order of magnitude larger than the one it was replacing, it has also provided a service almost ten times faster. A hierarchy of space and time territories emerges. As H. Simon (1988) noted, most natural and man-made systems are almost always hierarchical, i.e., they have boxes-within-boxes architecture. Transport systems portray a similar structure. By walking or using waterways a man can cover a mean circle of a few kilometers diameter in one hour - the size of a village or small town. A person travelling by rail or horse could travel more than a dozen kilometers in the same period a hundred years ago. The automobile and rapid rail systems offer a larger range - up to 100 km - and can thus effectively connect cities. Air travel extends the radius to almost 1,000 km. As connected territory increases so does travel; tangible goods transport and information flows increase in a unit of time. The next phase of expansion and improvement in transport and communication systems could reduce the vastness of our planet to one single functional system.

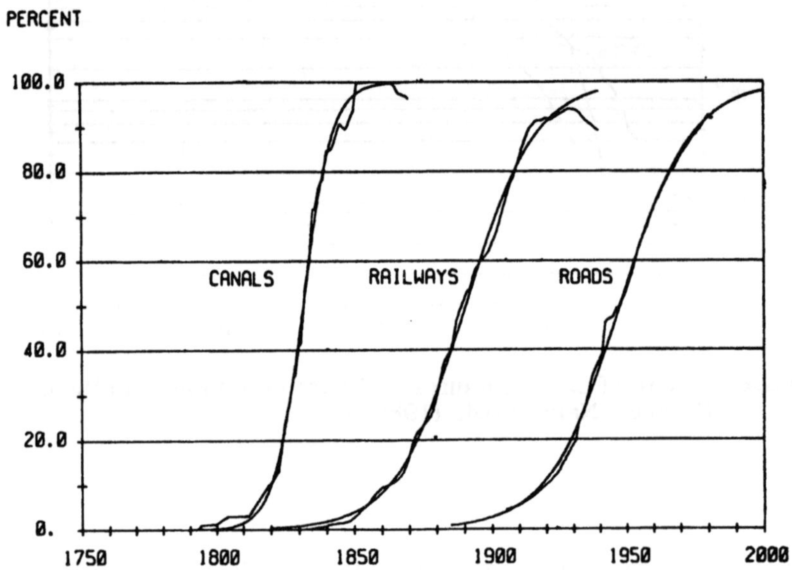

Figure 5.2 Diffusion of Canals, Railroads and Roads in the USA
Source: Nakicenovic (1988).

Figure 5.2 shows the expansion of three forms of infrastructure in the USA, normalized with respect to their respective saturation levels (by plotting the relative growth levels as a percentage of the saturation level). Thus, their succession can be described in terms of three S-shaped growth pulses that are given together with the estimated logistic curves. The three periods of saturation are separated in time by about six decades, while the respective introduction periods are much closer together. The development of canals, relative to the achieved saturation level, was much quicker than the expansion of railways and

roads. The time constant of growth, Δt is about 30 years for canals, 54 years for railroads, and 56 years forsurfaced roads. [1]
It is remarkable that despite these differences, saturation and the onset of decline of all three infrastructures coincides with the beginning of the prolonged recessions (in the 1870s, 1930s and 1980s)[2]. The life-cycles of various forms of infrastructure are indeed very long, spanning periods of a century until saturation is reached. The duration of senescence can be even longer. The most vital of the structures, however, are here to stay. Their immortality is marked by providing different services than those originally envisaged. More than a century after the canal era, the remaining inland waterways are now used for leisure activities, transporting low-value goods and irrigation. There are also more sails today than in the heyday of ocean clippers, but most of them are pleasure boats and do not carry a single commercial ton-kilometer.

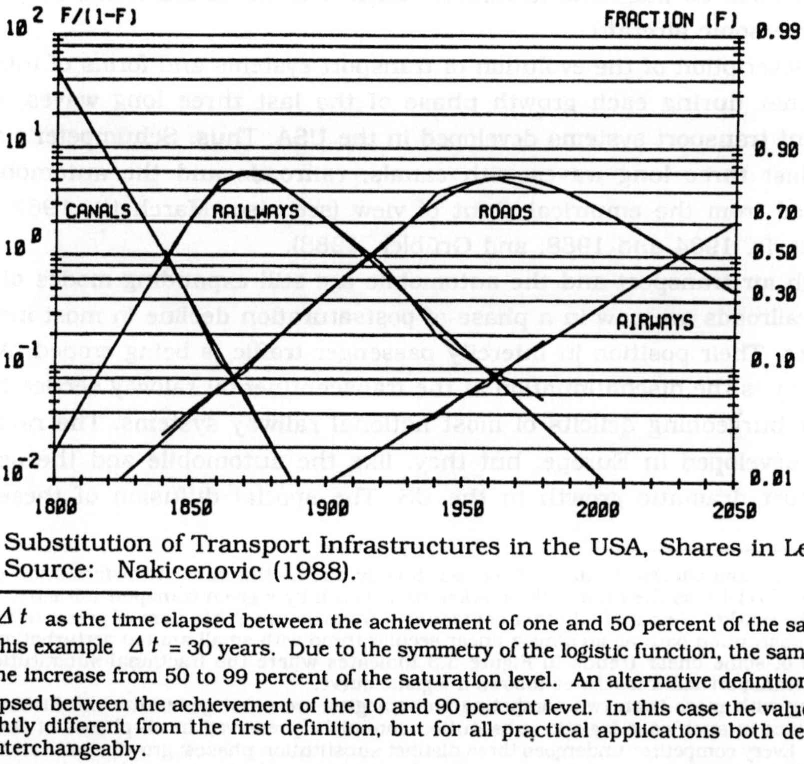

Figure 5.3 Substitution of Transport Infrastructures in the USA, Shares in Length
Source: Nakicenovic (1988).

[1]) We define Δt as the time elapsed between the achievement of one and 50 percent of the saturation level κ ,. In this example Δt = 30 years. Due to the symmetry of the logistic function, the same time is required for the increase from 50 to 99 percent of the saturation level. An alternative definition of Δt is the time elapsed between the achievement of the 10 and 90 percent level. In this case the value of Δt would be slightly different from the first definition, but for all practical applications both definitions can be used interchangeably.

[2]) The evolutionary path of successive replacements of traditional by new paradigms as seen in the diffusion of technologies and institutions, economic restructuring and transformations in social relations is captured in the Schumpeterian notion of long waves in economic development, i.e. the see-saw like pattern of the Kondratieff (1926) pulses experienced in the market economies during the last two centuries. Freeman (1983), Mensch (1975), Marchetti (1985), Nakicenovic (1984) and many others have subsequently extended both the conceptual and empirical description of long waves, diffusion, invention and innovation processes, albeit from different methodological and theoretical perspectives.For a comprehensive review of this research area, see van Duijn (1983), Vasko (1987) or Goldstein (1988).

Figure 5.3 shows the successive substitution of the three forms of transport infrastructure and the federal airways route miles. The shares of each infrastructure in the total length are plotted as the ratio of the share of one infrastructure divided by the sum total shares of all others on a semi-logarithmic scale.[1] This particular representation shows the relative importance of competing infrastructures and the dynamics of the evolution process during the last 180 years. In any given period, there is a clear market dominance (i.e., more than a 50 percent share) and at the same time simultaneous spread of transport activities over two or three different systems. Thus, while competing infrastructures are all used simultaneously, their mix changes over time leading to the expansion of the area connected in a unit of time. Projecting this competitive process into the future leads to the increasing importance of air routes and possible market dominance after the 2030s, notwithstanding the likelihood of a new competitor emerging during the coming decades, such as magnetic levitation (maglev) or advanced forms of air transport (e.g., hypersonic aircraft).

This description of the evolution of transport systems and forms of infrastructure shows that during each growth phase of the last three long waves, one of the important transport systems developed in the USA. Thus, Schumpeter's association of the last three long waves with canals, railroads and the automobile can be confirmed from the empirical point of view (see also, Marchetti, 1987 and 1988; Nakicenovic, 1984 and 1988; and Grübler, 1988).

Although air transport and the automobile are still expanding modes of passenger travel, railroads are now in a phase of postsaturation decline in most industrialized countries. Their position in intercity passenger traffic is being eroded. A symbol of this decay is the discontinuation of the transcontinental railway service in the USA, and the burgeoning deficits of most national railway systems. The railroads were largely developed in Europe, but they, like the automobile and the aircraft, had their most dramatic growth in the US. The spatial diffusion of these transport

---

1) The fractional shares, f , are not plotted directly but as the linear transformation of the logistic curve, i.e. f /(1-f), as the ratio of the market share taken by a given transport infrastructure over the sum of the market shares of all other competing infrastructures.This form of presentation reveals the logistic substitution path as an almost linear secular trend with small annual perturbations. Thus, the presence of some linear trends in Figure 5.3 indicates where the fractional substitution of various forms of transport infrastructures follows a logistic curve.

In dealing with more than two competing technologies, we must generalize the Fisher - Pry (1971) model, since in such case logistic substitution cannot be preserved in all phases of the substitution process. Every competitor undergoes three distinct substitution phases: growth, saturation and decline. This is illustrated by the substitution path of rail tracks, which curves through a maximum from increasing to declining market shares (see Figure 5.3). In the model of the substitution process, we assume that only one competitor is in the saturation phase at any given time, that declining technologies fade away steadily at logistic rates, and that new competitors enter the market and grow at logistic rates. As a result, the saturating technology is left with the residual market shares (i.e., the difference between 1 and the sum of fractional market shares of all other competitors) and is forced to follow a nonlogistic path that joins its period of growth to its subsequent period of decline. After the current, saturating competitor has reached a logistic rate of decline, the next oldest competitor enters its saturation phase and the process is repeated until all but the most recent competitor are in decline. A more comprehensive description of the model and assumptions is given in Nakicenovic (1979).

systems throughout Europe and North America has been treated in much greater detail in a recent study by Grübler (1988); suffice it here to mention that similar evolutionary changes in the development of transport infrastructures also occurred in the European countries. Even the planned economies have undergone the same evolutionary path (see Chapter 4).

5.3 PASSENGER TRAVEL

This irreversible process of evolutionary change, and the survival of the "fittest" in an ever changing environment that selects for speed, range and productivity, is nevertheless only a proxy for the real dynamics in the development of transportation systems; which should be measured in some common performance unit. Because transport systems provide a whole range of services, such a common descriptor is difficult to define. Furthermore, most transport services require a mix of different transport modes. This mix apparently changes in time. Two obvious choices for an appropriate indicator are ton- or person-kilometers per year. These units distinguish between freight and passenger services (although these are usually offered simultaneously) but they do not distinguish between short and long distances. Thus, there is no obvious shortcut and it appears necessary to analyze both passenger and freight separately for both short and long distances. Fortunately, the available historical data make it possible to reconstruct the dynamics of these substitution processes for at least some countries. As these results have been reported elsewhere (see for example, Nakicenovic, 1984 and 1988; Marchetti, 1987 and 1988; and Grübler, 1988), here we will briefly describe the modal split for long-distance passenger transport in the US and USSR, and then the evolution of freight transport in the FRG. These examples are representative of the developments in most industrialized countries and are given here to illustrate the convergence in the development of transport systems.

Figure 5.4 shows the substitution of different transportation modes in intercity (long-distance) passenger travel in the USA since 1950. By excluding urban and metropolitan transport, the competition for intercity passenger traffic is reduced to four significant transport modes: railways, buses, cars and airways. Figure 5.5 shows a similar substitution process in the USSR since 1920 (A. Grübler, 1988). The major competitors over this period were boats, railways, buses and airways; the automobile never gaining any significant importance in the USSR. Today, people travel by car and plane in the USA and by rail, bus and plane in the USSR. Comparisons of the two figures show that intercity passenger transport is in different phases of development in the two countries. Rail and bus are virtually extinct in the USA, while in the USSR they are still important means of transport.

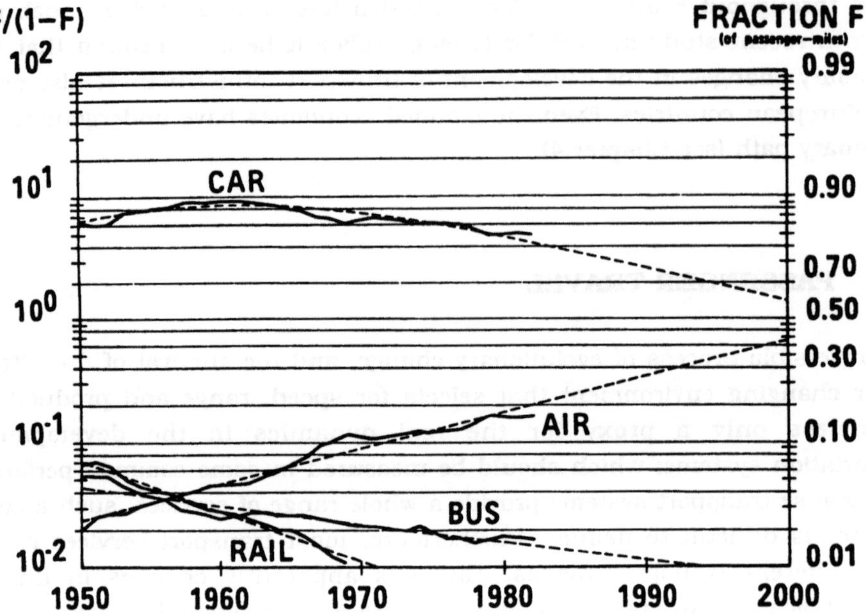

Figure 5.4 Substitution of Intercity Passenger Transport Modes in the USA, Shares in passenger-kilometers.
Source: Nakicenovic (1988).

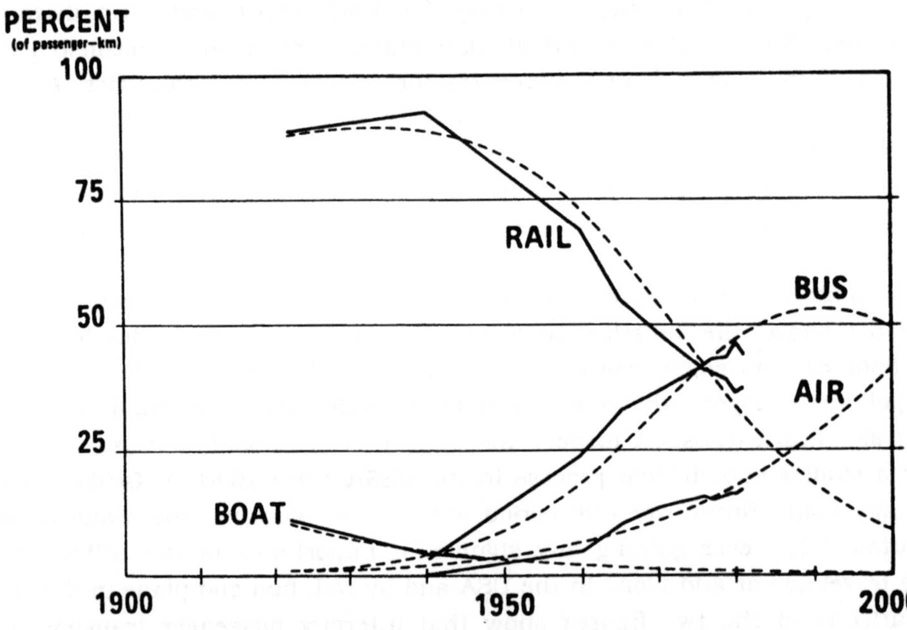

Figure 5.5 Substitution of Intercity Passenger Transport Modes in the USSR, Shares in passenger-kilometers.
Source: Grübler (1988).

Table 5.1 Time Budgets in 12 Countries, in Hours per Day

	Average Time per Person Spent on Activity						
	Household				Travel		
Country/city	Work	Children,..	Sleep	Leisure	Work	Non-work	Total
Belgium	4.38	5.65	8.35	4.73	0.40	0.53	0.93
Bulgaria, Kazanlik	6.05	6.04	6.97	3.55	0.68	0.80	1.48
Czechoslovakia, Olomouc	5.07	6.34	7.80	3.77	0.55	0.45	1.03
France, 6 cities	4.25	6.73	8.30	3.85	0.37	0.60	0.97
FRG, 10 districts	3.87	6.87	8.50	4.18	0.30	0.35	0.65
FRG, Osnabrück	3.63	6.60	8.34	4.68	0.27	0.70	0.97
GDR, Hoyerswerda	4.63	6.81	7.90	3.70	0.53	0.47	1.00
Hungary, Gyor	5.55	6.30	7.88	3.10	0.68	0.55	1.23
Peru, Lima Callao	3.57	5.97	8.28	4.68	0.62	0.88	1.50
Poland, Torun	4.97	5.92	7.78	4.10	0.62	0.68	1.30
USA, 44 cities	4.03	6.15	7.83	4.75	0.42	0.88	1.30
USSR, Pskow	5.65	5.43	7.70	3.77	0.55	0.92	1.47
Yugoslavia, Kragujevac	4.00	6.08	7.87	4.87	0.45	0.83	1.28

Note: rows may not sum to 24 hours due to rounding errors.

Source: Szalai (1972).

In the European countries the share of railways is still relatively large but smaller than in the USSR. The substitution dynamics indicate that by the end of the century airways may become the dominant form of intercity travel in both countries. The evolutionary paths appear to be convergent. In both countries road transport currently has the largest share, albeit by individual automobile in one case and by bus in the other. What is important is that the average choice of different modes of passenger travel changes consistently in both countries and favors faster and more productive systems.

Zahavi (1979) has shown that travelling is optimized under the constraints of individual time and family income budgets. On average, roughly one hour a day and 15 percent of disposable family income is allocated to travel. Table 5.1 illustrates the average time budgets of the inhabitants of 12 cities throughout the world, including both industrialized and developing countries (Szalai, 1972). On average, the time spent for travelling (work and non-work) is roughly one and a half hours a day. Marchetti (1987) argues that the allocation of time and money is made so as to maximize distance. In a given situation, each individual will make different choices consistent with this hypothesis, but on average the modal split will change as income increases, despite the assumed invariance of travel time budgets. Incomes have increased and the cost of travel has decreased in real terms, leading to an increase in the volume of travel (passenger-kilometers) and the range in a given unit of time. Over time larger shares of travel are thus allocateOd to faster modes of travel. The slower modes recede to service fewer people over shorter distances (fewer passenger-kilometers) and low value segments of freight transport. Their shares in passenger transport decline, resulting in the rationalization of the respective infrastructures. The least productive links are decommissioned and the infrastructure declines. The diffusion of more productive and faster means of transport leads to increased range and connects even larger areas into one single complex. During the last two centuries villages have merged into towns, towns into cities and cities into metropolitan areas. Large urban corridors have evolved throughout the world, some approaching one hundred million inhabitants.

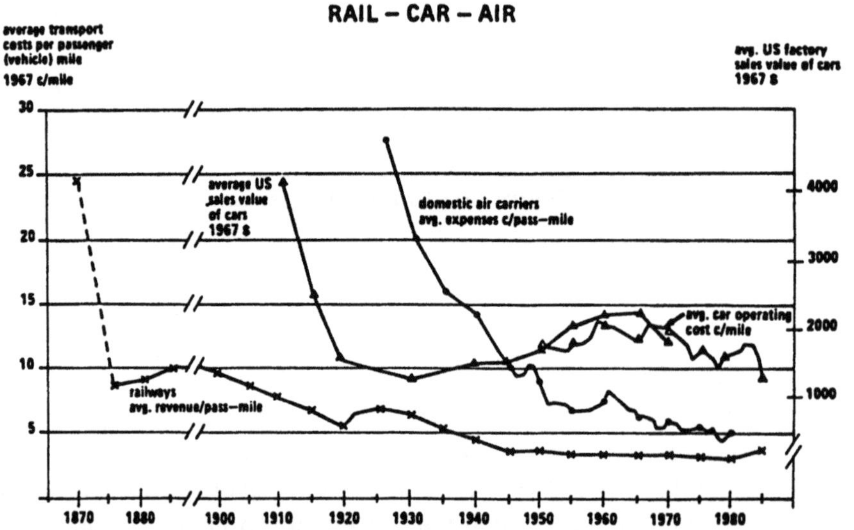

Figure 5.6 Indicators for Average Passenger Transport Costs in the USA
Source: Grübler (1988).

Figure 5.6 shows the dramatic decrease in the costs of passenger-miles for railways, automobiles and air travel in the USA (Grübler, 1988). Over time, costs have decreased for each mode of transport, but during the period of greatest expansion in the shares of the respective infrastructure, the relative costs were the highest. The average revenue passenger-mile costs in the USA were very high during the expansion of railways (prior to the 1870s). The automobile replaced horse-driven vehicles in the first decades of this century (see Nakicenovic, 1987), during the period of relatively high operating costs of cars (measured in terms of vehicle sales value up to the 1950s and operating costs thereafter); air travel increased dramatically while the costs were the highest of all competing modes. These two developments would appear to be inconsistent from a static equilibrium point of view. However, allowing for increases in disposable income since the onset of the industrial revolution, the maximization of travel range would lead to the observed changes in the modal split as reflected in the lengths of various forms of transport infrastructure, the structure of intercity travel and transport costs.

5.4 FREIGHT TRANSPORT

Time is a valuable resource. While the possibility of constant travel time allocation would tend to increase the demand for faster modes of passenger transport subject to money budgets, the changes in the modal split of freight transport are caused by different driving forces. The transportation of tangible goods imposes more stringent requirements with increasing goods-value (value per unit weight or volume). Long transport times require larger inventories and correspondingly higher capital expenditure. Furthermore, the risk of loss or damage also tends to increase with the exposure time. All told, more valuable goods are transported by faster transport modes.

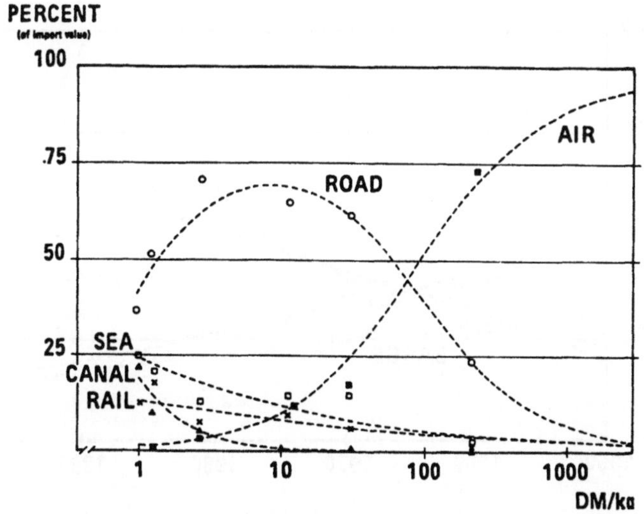

Figure 5.7 Substitution of Freight Transport Modes in the FRG, Shares in ton-km
Source: Grübler (1988).

Figure 5.7 illustrates this development for domestic routes in the FRG. Development in other industrial countries reflects the same historical pattern. The share of waterways and rail in goods transport is decreasing and road dominance increasing over time. Here, higher quality of service, smaller batch sizes, faster and more accurate deliveries make the difference. Despite higher ton-kilometer charges, the total incurred transport costs can be lower due to the shorter duration of transport operations and higher quality of services. With the introduction of flexible manufacturing, larger worldwide transport of perishable goods and fashionable garments, the air-road systems are increasing their competitiveness. The increasing goods value of tangible goods and decreasing materials intensiveness tends to generate higher value added on the most efficient routes. Express business in general is composed of millions of small, high-value shipments. In fact, F.W. Smith (1987), CEO of Federal Express, disclosed that supersonic transport of freight would generate almost three times the revenue than supersonic Concorde does on transatlantic routes with passengers on board. In the past, the first user of aircraft was high priority freight. Mail services preceded passenger travel in the pioneering days of aviation. Grübler (1988) has shown that new means of transport enter the highest quality market segments first, prior to attracting bulk and low-value goods. The converse is also true, outdated modes of transport lose the most precious market segments first.

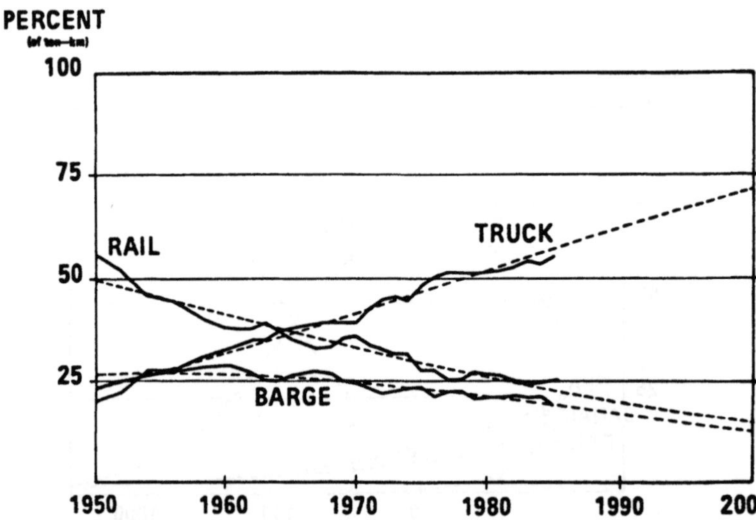

Figure 5.8 Substitution of Transport Modes in the Import Value of Manufactured Goods Versus Goods-Value in the FRG, Shares in DM/kg
Source: Grübler (1988).

Figure 5.8 shows a snapshot in time of imports of manufactured goods into the FRG as a function of goods-value. This figure illustrates that the most valuable goods are transported by more efficient and faster means of transport. Basic materials such as coal, gravel, scrap and ores in the value range of a few DM per kilogram are mostly transported by sea, canal and rail. But even in this lowest segment, trucks are more competitive and are the dominant transport mode up to about 100 DM per kilogram. Most of the manufactured goods such as automobiles and machine tools fall into the range below this threshold. Electronics, computer and precision instruments are usually freighted by air. The highest value manufactured goods that are exclusively transported by air are aerospace products and aircraft themselves.

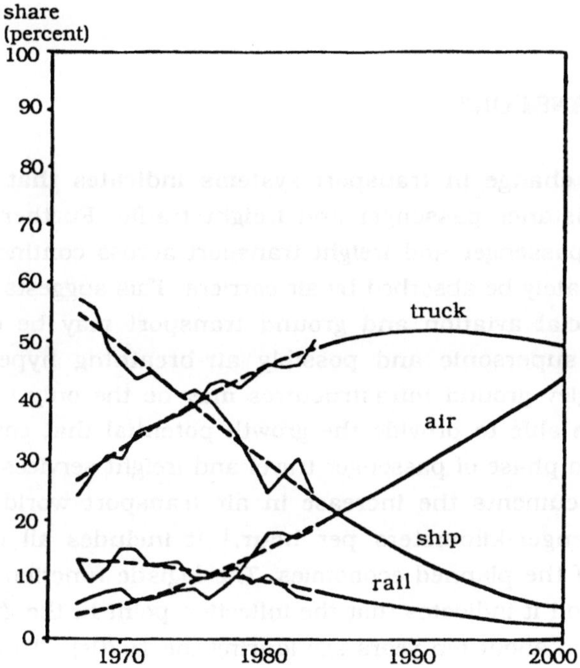

Figure 5.9 Substitution of Transport Modes in the Export Value of Manufactured Goods in Sweden, Shares in SKr
Source: Snickars (1987).

Recently, General Motors started a 5,000 kilometers "production line" connected by airfreight. Cadillac car bodies are assembled at Pininfarina in Turin, Italy, and transported by Lufthansa and Alitalia B-747 cargo aircraft to Detroit for final assembly of engine, power train and electronics. This airlift assembly line is apparently economic compared to all the direct and indirect costs of potential damage risks and the production inventories that would be locked in ocean freighters for a few weeks. On a more speculative note, the further dematerialization of manufactured goods and the increase of value through software and information content would tend to further increase the share of air transport in freight. Co-location of production facilities and services close to airports may proceed in a similar way as industrial activities which concentrated along previous transport infrastructures: railways and later roads.

Figure 5.9 confirms this overall trend for Sweden (Snickars, 1987; see also Chapters 4 and 6). If the total value of imported goods is used as an indicator of freight allocation among disposable transport modes instead of ton-kilometers, the figure shows that airfreight already controls a large and increasing share of imported and exported goods.

5.5 AIR TRANSPORT

The evolutionary change in transport systems indicates that faster modes are favored in long-distance passenger and freight traffic. Furthermore, the bulk of future growth in passenger and freight transport across continents and the globe would most adequately be absorbed by air carriers. This suggests that a new growth pulse in commercial aviation and ground transport may be due in about two decades. Cruise supersonic and possibly air-breathing hypersonic transports connected to maglev ground infrastructures may be the only technologies on the horizon which are able to provide the growth potential that could be required in the next expansion phase of passenger travel and freight services.

Figure 5.10 documents the increase in air transport worldwide measured in millions of passenger-kilometers per hour.[1] It includes all carrier operations, including those of the planned economies. The logistic function has been fitted to the actual data, and it indicates that the inflection point in the growth of air carrier operations occurred about ten years ago (during the 1970s). Thus, after a period of rapid exponential growth, about one doubling is left until the estimated saturation level is achieved after the year 2000.

Perhaps the most interesting result is that it took about 30 years for world air transport to reach the inflection point (about half of the estimated saturation level)

---

[1]) Historical data and a fitted logistic curve are transformed as $x/(\kappa - x)$, where x denotes the actual volume of all operations in a given year and κ the estimated saturation level. The data and the estimated logistic trend are plotted as fractional shares of the saturation level, $f=x/\kappa$, which simplifies the transformation to $f/(1-f)$, the level of relative growth achieved divided by the remaining potential. Transformed in this way, the data appear to be on a straight line, which is the estimated logistic function. Without this transformation, the data and the trend curve would portray the same S-shaped growth as shown for the three transport infrastructures in Figure 5.2: canals, railways and surfaced roads.

and that within two decades the saturation level will be reached. This raises a crucial question: What could happen after such a saturation?

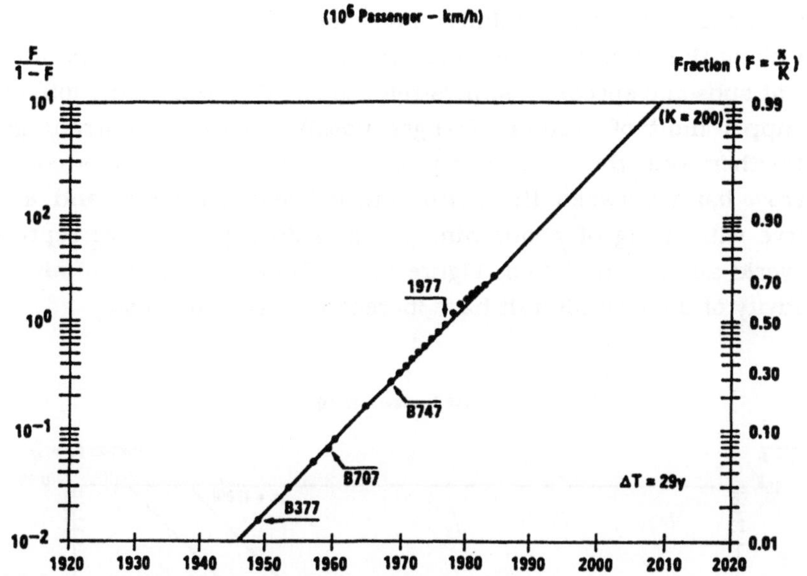

Figure 5.10 Air Passenger Transport Worldwide, Logistic Plot, pass-km/h Source: Nakicenovic (1988).

The global fleet of commercial aircraft can be described in a number of ways. An obvious descriptor of the fleet is the number of commercial aircraft in operation worldwide. This number increased from about 3,000 in the 1950s to almost 10,000 in the 1980s. During the same time, however, the performance or carrying capacity and speed of aircraft increased by about two orders of magnitude. Thus, the size of the fleet is not the most important descriptor, because much of the traffic is allocated to the most productive aircraft operating between the large hub airports while other aircraft constitute the feeder and distribution system for destinations with lower traffic volume.

Figure 5.11 shows the improvement in carrying capacity and speed (often called productivity) measured in passenger-kilometers per hour. Each point on the graph indicates the performance of a given aircraft when used in commercial operations for the first time. The DC-3 was introduced in 1935 with a performance of about 7,400 pass-km/h (21 passengers at 350 km/h); the B-747 was introduced in 1969 with a performance of about 500,000 pass-km/h (500 passengers at 1,000 km/h). The largest planned B-747 series could carry almost 700 passengers. The B-747 family is therefore about one hundred times as productive as the DC-3 50 years ago. The upper curve in Figure 5.11 represents a kind of performance feasibility frontier for passenger transport. Thus, at any given time there appears to be only one appropriate productivity specification for long-range (most productive) passenger planes. All commercially successful long range transports were introduced along this frontier. The saturation level is estimated to be about 1,200 passengers at subsonic speeds. A superstretched B-747 could do the job, but that is the likely upper limit of gradual changes possible with the current generation aircraft. Furthermore, the productivity of all passenger aircraft is confined to a rather narrow band between the performance feasibility curve and a "parallel" logistic curve with a lag of about nine years. This logistic curve represents the growth of world air transport from Figure 5.11. Thus, the growth of air travel and the productivity of the best aircraft have increased at similar rates.

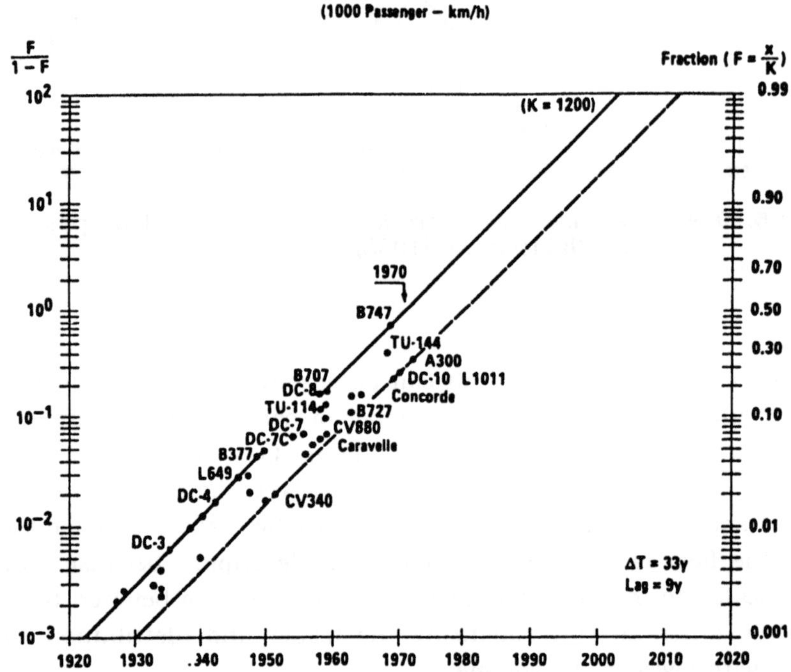

Figure 5.11 Passenger Aircraft Performance, Logistic Plot, pass-km/h
Source: Nakicenovic (1988).

The fastest airliner, Concorde, and the most productive jet transport, Boeing 747, flew for the first time twenty years ago. Ever since, the volume of all airline operations has continued to increase while the productivity of the B-747 increased marginally through stretching. Most of the subsequent technological developments have been focused on gradual improvements such as noise reduction and the improvement of fuel efficiency. Unavoidably, this leads to an impression of diminishing returns in the further advancement of airliners and, consequently, the possibility of an approaching saturation in commercial air transport. An alternative view would be to expect an increase in the number or capacity of aircraft in order to keep up with increases in the volume of traffic.

Our analysis of the development of various forms of transport infrastructure and passenger travel indicates that large increases in the volume of operations are indeed likely, as air transport continues to take larger shares of traffic and total passenger-kilometers traveled worldwide increase with further economic growth. The likelihood of hyperjumbos occupied by a few thousand passengers is slim, considering current airport overcongestion and the many inconveniences associated with an increasing size of aircraft. A third way out of this stalemate could be enhanced productivity of the fleet by increasing aircraft speed, which would represent a revolutionary change in technology and a whole host of possible positive and negative externalities.

The third alternative in providing for further increases in commercial air transport well into the next century could materialize if the renewed research programs in advanced, high-speed airliners by the aerospace communities on both sides of the Atlantic bear fruit. Recently, plans were announced to study, and eventually develop, supersonic (SST) and hypersonic transports (HST). The most daring of the proposed designs could fulfill even more ambitious objectives and lead to the development of an air-breathing spaceplane.

Assuming that such a development is technologically possible and economically viable, it poses a number of fundamental questions associated with the expected side-effects of a large fleet of hypersonic passenger aircraft. Certainly, the noise problem must be overcome by reducing the inherent noise profiles and removing the hubs and routes for such aircraft from populated areas. A more important concern, however, is the possible adverse effect of such a fleet on the chemistry of the upper atmosphere. In view of the observed ozone depletion of the stratosphere, such concerns are particularly timely while the design and configuration of proposed hypersonic transports are still in an early development phase. Aircraft capable of operating at high Mach numbers would be powered by methane or hydrogen. At hypersonic speeds in excess of Mach 5, hydrogen would be the fuel of choice since, among other things, it would create the most environmentally benign emissions. On the other hand, it would result in at least large quantities of water vapor at high altitudes. Recent studies of the Antarctic "ozone hole" have clearly shown that the stratosphere can be very sensitive to an increased concentration of chlorofluorocarbon gases in the atmosphere, and that this sensitivity may be strongly enhanced if more water vapour is added to the stratosphere. Other possible effects should also be considered, such as higher water vapour concentrations along the flight routes which, directly or indirectly through the formation of contrail cirrus clouds, can lead to enhanced "greenhouse" heating of the earth.

The possibility of adverse effects from a continued increase in transport operations worldwide warrants further studies. If they could be overcome, the maximization of range through productivity increases and speed would be consistent with the observed evolution of transport systems since the onset of the industrial revolution. Marchetti (1987) postulated that this evolutionary process will be complete when any place on the whole earth becomes reachable within an hour. A gigantic "ecumenopolis" could be serviced by highly productive aircraft and maglev infrastructures, providing that such developments become socially acceptable and that the current prolonged phase of economic restructuring is followed by renewed growth leading to unprecedented levels of welfare. An indicative precursor of this development is the deployment of B-747 aircraft for short distance commuting along the Tokyo-Osaka corridor in addition to the Shinkansen rapid rail transport. In the not-too-distant future, a maglev link could make the two metropolitan areas into a truly integrated hypercity that could be traversed within an hour. The off-shore Kansai airport could then provide an extremely productive hub for long distance, transcontinental travel.

5.6 INFORMATION AND COMMUNICATION

The evolutionary development process of transport systems over the last two centuries indicates that the replacement of older forms of infrastructure and transport modes by newer ones has not only improved the productivity, range, speed and quality of services, but at the same time has resulted in an increasing complexity of systems and diversification. Canals were used for most transport needs; they transported people and sources of energy (flotation of timber), provided power (uses of hydropower for direct mechanical drives), basic materials and finished products. Today there are a number of very specialized infrastructures, carriers and services. Examples range from pipelines to air courier services. A similar change has occurred with information and communication technologies. Postal services were among the first information and communication systems available to the general public. Most of the mail was transported in the same way as passengers and freight, although it often had a higher priority. The telegraph, and later telephone, represented a new development and were a gigantic quantum leap in increasing the speed of communications almost to an instantaneous level compared to mail. They also required dedicated infrastructures.

Most of the modern information and telecommunication systems are also based on dedicated infrastructures. The bifurcation of transport and communication systems has remained apparently permanent since the introduction of the telegraph. Even in those early days, however, the two infrastructures (railroads and telegraph) functioned in a symbiotic relation to each other. They enhanced each other. Nevertheless, when the telephone was introduced there was hope that it would replace some of the personal travel requirements, especially in rural areas with low population densities. This apparently did not occur. Today there is considerable speculation as to whether advanced information and communication technologies will replace some of the face-to-face contacts that require personal travel. Historical experience does not suggest such a possibility. As a social animal, man requires personal contact and even the most advanced technologies do not communicate all the signals exchanged during face-to-face intercourse.

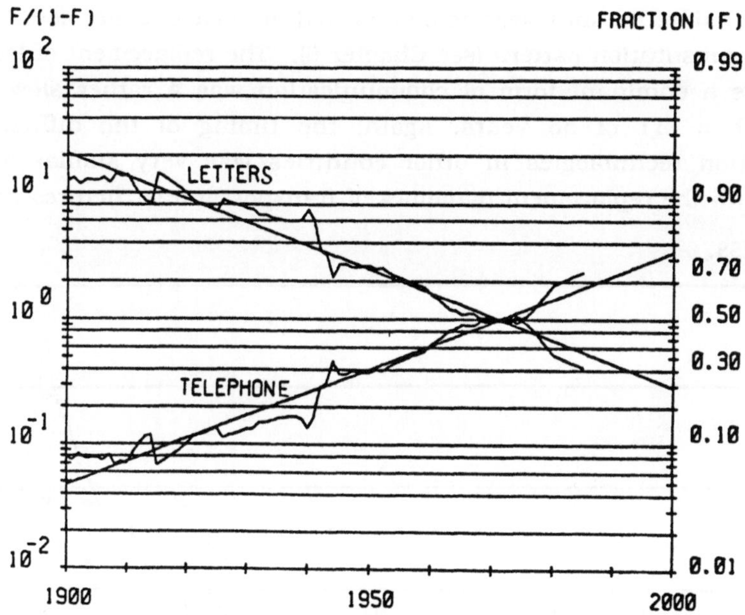

Figure 5.12 Substitution of Letters by Telephone in France, Shares in Total Messages
Source: Grübler (1988).

The validity of this hypothesis is difficult to measure empirically. Quantitative assessment of transport operations is hampered by the difficulties in obtaining relevant data because it is not always certain which variables are the most appropriate ones to measure. In order to describe the evolution of transport systems, the length of various forms of infrastructure and the volume of operations were used. In the case of information and communication systems the measurement is even more difficult, not only because much of the information is redundant, but because the information content relevant for comparing the performance of competing systems cannot be expressed in bits or bauds. Instead, the number of messages exchanged would be a more appropriate indicator.

As a proxy for the performance of information and communication systems, Figure 5.12 shows the shares of total messages transferred by postal services and telephone, namely the number of letters and phone calls per year (Grübler, 1988). The share of total messages represented by telegrams, and later telexes, was always rather low, and never exceeded a one percent share. Newer communication channels such as facsimile are not included due to the lack of data, but their share is no doubt increasing. Figure 5.12 clearly indicates that the substitution of communication systems can be described in the same way as transport systems. The process shows the enormous impact of the telephone, which, for example, accounts for half of the messages communicated in France. Other countries portray an identical substitution pattern (see Chapter 6). The replacement of letters by the telephone as a dominant form of communication was a rather slow process in France, with a Δt of 53 years. Again, the timing of the diffusion of new communication technologies in other countries was very similar and can be compared with the replacement dynamics of transport infrastructures and services.

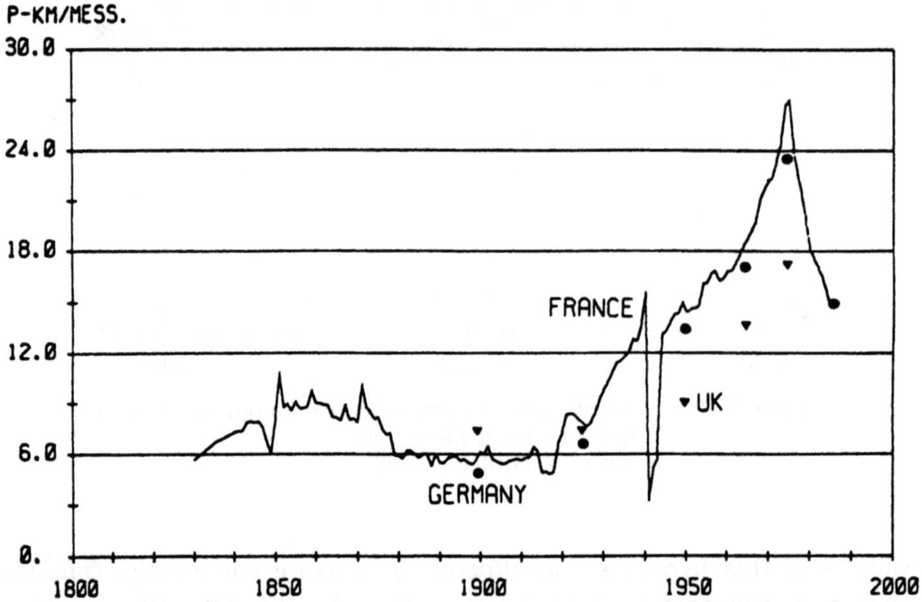

Figure 5.13 Ratio of Passenger Travel to Messages Sent in France, the UK and the FRG, pass-km/messages
Source: Grübler (1988).

Building on this tenuous and tentative evidence that the evolution of transport and communication systems follow similar structural changes, a comparison of the improvements in performance of the two systems should reveal to what extent the substitution effects can be determined and to what degree the two systems evolved in unison. Figure 5.13 gives the ratio of total passenger-kilometers of personal travel divided by all messages transmitted through communication systems for France, the UK and the FRG (Germany before the 1940s). Thus, the ratio reflects to some degree the relative increases in communication through physical travel (e.g. face-to-face) compared with communication through information channels.

As can be seen from this figure, the ratio increased substantially from the early 1920s until the 1970s, when a trend reversal occurred. One probable explanation is the enormous increase in mobility through the diffusion of the automobile relative to the more modest increase in information flows due to the telephone. Today, communication via information channels is increasing more rapidly. Certainly, a number of business trips can be substituted by new and more productive communication technologies. Advanced electronic systems as well as the emerging photonic technologies will increase the relative information capacity of transmission channels by a few orders of magnitude (e.g. see National Research Council, 1988). New communication services are also expanding rapidly, such as the use of facsimile integrated into personal computers or mobile telephones. At the same time, the new communication technologies may in turn induce additional travel. It is very likely that reductions in the amount of business and daily errand-related travel could lead to more time being devoted to leisure and holiday travel. In addition, more effective information and communication technologies may require more efficient travel. As the distance between individuals who are in a constant and effective communication network expands, more extensive and faster travel will be required for those occasions when personal contact is absolutely necessary. Thus, while it is true that the growth of transport and communication did not evolve in precise unison, they do enhance each other and show many parallels. This is an empirical indication that they are complementary services rather than substitutable ones. A more comprehensive examination of this dynamic process is recorded in the next chapter.

The above conclusion refers only to the aggregate level and does not hold for each individual. For society as a whole, however, travel and the transportation of information and tangible goods are all different dimensions of a vast network coordinating activities across time and space, and providing services and goods that cannot be generated simultaneously at all locations. In the long run the advent of more effective and productive information technologies will be accompanied by improvements in the performance of transport systems. More information will lead to more transport and vice versa. In effect, communication networks are already global whereas travel and the transport of goods is still rather slow over intercontinental distances. Today courier and express services represent integrated transport systems that can handle millions of shipments around the world, each with the individual priority, speed and flexibility that almost instantaneous communication worldwide has made possible and increasingly requires.

5.7 CONCLUSION

Obviously, the above description of these complex systems is incomplete. Certainly, there are other illuminating ways of describing the dynamics of transport and communication systems. The intriguing aspect that emerges from the development of transportation and communication systems is that they appear to be interwoven with regular features reflected in the pattern of substitution dynamics over a period of about two centuries. The analysis suggests that a new growth pulse in commercial aviation and ground transport may be due in about two decades. Cruise supersonic and possibly air-breathing hypersonic transports connected to maglev ground infrastructures might be the only technologies on the horizon to provide the required growth potential in the next expansion phase of passenger travel and freight services.

The growth potential of communication technologies appears to be even greater. Historically, the expansion of communications lagged behind developments in travel and trade. Since transport and communication are complementary human needs, there is an opportunity for improving information flows to historical parity with the exchange in tangible goods and travel. A number of critical emerging technologies hold promise for large growth potential. They include the advances in high-temperature superconductivity - much discussed over the last three years - that may revolutionize electronic communication systems over the next decades as well as applications of photonic technologies to information processing, storage and communications. Further developments in these technologies will improve the bandwidth, speed, accuracy and distance encompassed by future telecommunication systems, information processing and control applications.

K. Kobayashi, CEO of NEC, has been saying unequivocally that the great prosperity in the USA could be linked directly to the unprecedented development of its advanced communication and transportation systems. We are on the brink of the next quantum leap in global mobility. A renewed phase of prosperity and growth will also demand new transport and communication technologies, as mobility, information and material wellbeing are symbiotic ingredients of modern societies. The key-technologies are emerging and will most certainly be commercialized. This development will, however, not proceed equally in all corners of the world. To that extent it will also affect the relative competitiveness of various places, strengthening the leaders and weakening those who fall behind. It is not certain what the technoeconomic paradigms of tomorrow will be, but even today effective transport and communication systems bring decisive advantages. To be sure, it will not be sufficient to develop transport and communication technologies as a guarantee for the future. Rather, they will be required in a renewed period of growth and restructuring as new sectors emerge.

REFERENCES

van Duijn, J. (1983) *The Long Wave in Economic Life* . London: Allen and Unwin.

Fisher, J.C., and Pry, R.H. (1971) A Simple Substitution Model of Technological Change. *Technological Forecasting and Social Change*, 3:75-78.

Freeman, C. (Ed.) (1983) *Long Waves in the World Economy*. London: Butterworths.

Goldstein, J.S. (1988) *Long Cycles: Prosperity and War in the Modern Age*. New Haven & London: Yale University Press.

Grübler, A. (1988) *Rise and Fall of Infrastructures, Dynamics of Evolution and Technological Change in Transport*. Ph.D. Dissertation, Vienna: Technische Universität.

Kondratieff, N.D. (1926) Die langen Wellen der Konjunktur. *Archiv für Sozialwissenschaft und Sozialpolitik* , Tübingen: Verlag Von J.C.B. Mohr, Band 56:573-609.

Marchetti, C. (1985) Swings, Cycles and the Global Economy. *New Scientist*, 1454:12-15.

Marchetti, C. (1987) On Transport in Europe: The Last 50 Years and the Next 20. in *Transport in Europa* , First Forum on Future European Transport, Universität der Bundeswehr München, September 14 - 16 1987. Ludwig-Bölkow-Stiftung, Ottobrunn/München, and SESTA, Paris.

Marchetti, C. (1988) *Building Bridges and Tunnels: The Effects on the Evolution of Traffic*. SR-88-01, Laxenburg, Austria: International Institute for Applied Systems Analysis.

Marchetti, C. (1988) *Infrastructures for Movement: Past and Future* , in J.H. Ausubel and R. Herman (Eds.), *Cities and Their Vital Systems. Infrastructure Past, Present, and Future* , Washington D.C.: National Academy Press.

Mensch, G. (1975) *Das Technologische Patt* , Frankfurt am Main: Unschau Verlag.

Modis, T. and Debecker, A. (1988) Innovation in the Computer Industry. *Technological Forecasting and Social Change* , Vol. 33, 3:267-278.

Nakicenovic, N. (1979) *Software Package for the Logistic Substitution Model* . RR-79-13, Laxenburg, Austria: International Institute for Applied Systems Analysis.

Nakicenovic, N. (1984) *Growth to Limits, Long Waves and the Dynamics of Technology.* Laxenburg, Austria: International Institute for Applied Systems Analysis.

Nakicenovic, N. (1988) *Dynamics of Change and Long Waves.* WP-88-74, Laxenburg, Austria: International Institute for Applied Systems Analysis.

Nakicenovic, N. (1988) Dynamics and Replacement of U.S. Transport Infrastructures, in J.H. Ausubel and R. Herman (Eds.), *Cities and Their Vital Systems. Infrastructure Past, Present, and Future* , Washington D.C.: National Academy Press.

National Research Council (1988) *Photonics, Maintaining Competitiveness in the Information Era* , Panel on Photonics Science and Technology Assessment, Washington D.C.: National Academy Press.

Schumpeter, J.A. (1935) The Analysis of Economic Change. *Review of Economics and Statistics*, 17:2-10.

Schumpeter, J.A. (1939) *Business Cycles, A Theoretical, Historical and Statistical Analysis of the Capitalist Process*, Volume I and II, New York: McGraw-Hill.

Simon H. (1988) *Prediction and Prescription in System Modeling.* Paper presented at IIASA Conference 1988; IIASA: Perspectives and Future, 14-15 June 1988, Laxenburg, Austria: International Institute for Applied Systems Analysis.

Smith, F.W. (1987) Practical Applications of Hypersonic Flight... Possibilities for Air Express, in J.P. Loomis (Ed.) *High Speed Commercial Flight, The Coming Era.* Proceedings of the First High Speed Commercial Flight Symposium, Columbus, Ohio, October 22-23, 1986. Columbus, Richland: Battelle Press.

Snickars, F. (1987) The Transportation Sector in the Communications Society: Some Analytical Observations, in P. Nijkamp and S. Reichmann (Eds.), *Transportation Planning in a Changing World.* Aldershot, Hampshire: Gower.

Szalai, A. (1972) *The Use of Time, Daily Activities of Urban and Suburban Populations in Twelve Countries.* Mouton: The Hague, Paris.

Vasko, T. (Ed.) (1987) *The Long-Wave Debate.* Selected Papers from a IIASA International Meeting on Long-Term Fluctuations in Economic Growth: Their Causes and Consequences, Weimar, GDR, June 10-14 1985. Berlin: Springer Verlag.

Zahavi, Y. (1979) *The "UMOT" Project.* Report DOT-RSPA-DPB-20-79-3, prepared for U.S. Department of Transportation, and Ministry of Transport, FRG. Springfield, Virginia: NTIS.

Chapter 6

THE FUTURE OF TRANSPORT AND INTERFACE COMMUNICATION: Debating the Scope for Substitution Growth

David F. Batten
University of Umeå, Sweden

6.1 INTRODUCTION

Transport and telecommunications are complementary technologies, jointly contributing to the efficient functioning of our economic and social systems. Yet, as means of interaction, they are also competitors. Greater use of telecommunications has partly replaced the need for travel in certain situations.

This chapter debates the future relationship of physical transport and interface communication. By necessity, the arguments presented on both sides are selective rather than exhaustive. A documented case for substitution growth is presented first, building upon observed instances of substitution and a report tabled at a recent European Conference of the Ministers of Transport. A documented case against substitution growth is then put forward, relying mainly on the extensive body of historical evidence demonstrating the complementary relationship between the two technologies. Summary arguments are delivered before the reader is invited to consider his or her own verdict.

To help the reader serve as a diligent juror, two additional criteria are proposed in order to scrutinize the merits and demerits of each case: a simple economic principle advocated by Dupuit almost 150 years ago, and the classical notion of "space-time prisms" advocated by Hägerstrand more recently. In the final analysis, the author´s position is portrayed as being little more than the same kind of educated guess that the typical reader will need to exercise. Thus this chapter illustrates how some of our agents of science, in this case transport and communication scientists, typically arrive at their view of the future, and that this view is constantly being challenged and updated in an ongoing clash of fact, theory, opinion and personality.

6.2 SUBSTITUTION ANALYSIS

Competitive forces have always exerted a powerful influence on the demand for various modes of transportation and communication. For example, an evenly paced and surprisingly predictable pattern of competition has been going on this century between the two principal modes of communication - the telephone call and the letter (see also Chapters 4 and 5). The trajectories of substitution for Sweden's two message markets (domestic and international) are depicted in Figure 6.1. Both the winner and the loser follow an S-shaped curve. A similar switch to the telephone is occurring throughout (Western) Europe. Table 6.1 records data describing the stage reached and the speed of this evolutionary process for ten European nations. This type of competitive substitution favouring the telephone call at the expense of the letter has been occurring throughout the industrialized world. In each case the pattern has been regular and predictable.

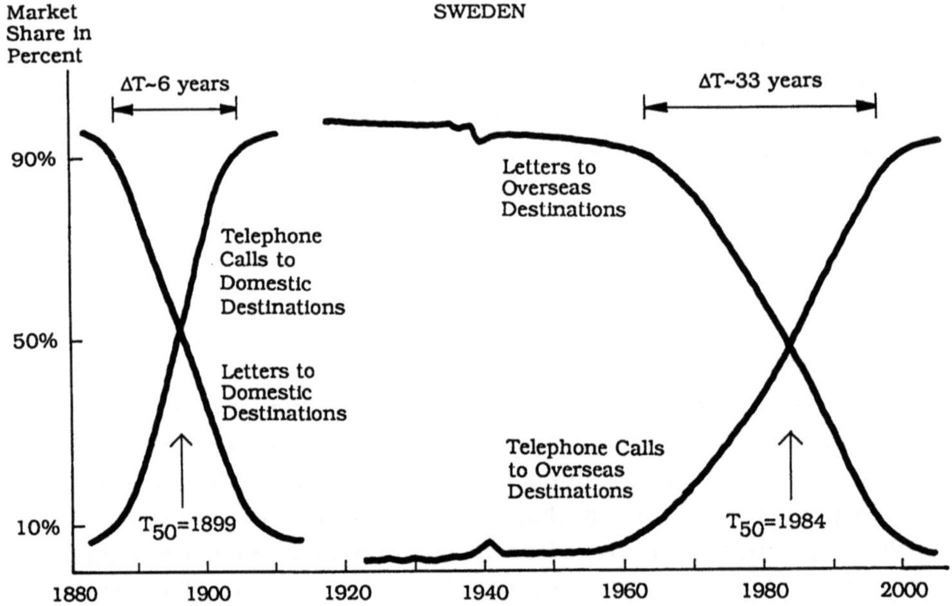

Figure 6.1 Substitution of Telephone Traffic for Letters in the Swedish Domestic and International Message Markets

Table 6.1 The Evolution of Telephone and Letter Traffic within European Nations

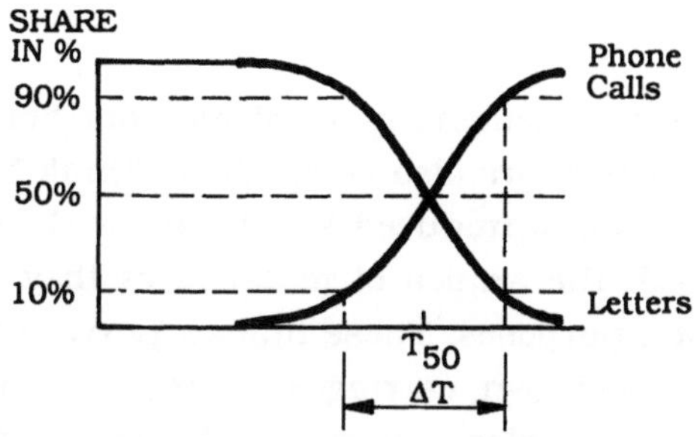

NATION	T_{50}	ΔT
Sweden	1899	6
Denmark	1909	15
Hungary	1956	15
Portugal	1961	55
Finland	1964	40
Switzerland	1967	70
W. Germany	1971	80
U.K.	1972	65
France	1972	70
Belgium	2015*)	110

*) This year is a forecast

We can observe a strikingly similar trajectory in the chronological development of infrastructure for transportation (see Figures 4.1 and 5.3). In this case, there are four competing modes which have been monitored by their length. Substitution between these modes is depicted in relative terms (i.e. as each mode's share of the total length of all four types of infrastructure). A traveler or a ton of goods can make a trip partly by road, partly by rail, and partly by air. In fact, to a large extent transport has always consisted of a combination of solutions. But the replacement sequence is clearly evident: canals → railways → paved roads → air networks. In the final analysis, air traffic is destined to increase by an order of magnitude even without taking the evolution of <u>*total*</u> traffic into account.

One can find numerous examples of international markets where competition between newer and older techniques, products, production systems or infrastructure exhibits a very similar replacement history. Market penetration, product cycles and processes of technological substitution seem to proceed at an even (though different) pace along a regular time trajectory. Among others, Fisher and Pry (1971), Peterka (1977), Marchetti (1987) and Nakicenovic (1987), have studied more than 300 competitive situations within the field of transportation and energy - all confirming the resemblance of the substitution process to biological (eg. Lotka-Volterra) models of competition. The logistic curve generally provides a good fit to these data. Each equation has only two parameters to be fitted: a time shift and a rate parameter. A short summary of the mathematics corresponding to this type of curve-fitting exercise appears in the Appendix.

6.3 THE CASE FOR SUBSTITUTION GROWTH

6.3.1 Introduction

Transport and telecommunications maybe complementary services from both a technological and a functional viewpoint, but they are also competitive. Think back to the previous century when the telephone was introduced by Alexander Graeme Bell. It enabled many firms to dispense with the armies of messengers they had commandeered previously for communication purposes. These human purveyors of information had relied upon legwork, horse-drawn carriages, boats and other primitive means of physical transport. Once the telephone replaced them, part of the demand for transport services was lost. Telecommunications could thus be seen as a substitute for some travel.

6.3.2 Travel versus Interface Communication

Before we can explore competitive trends between transport and communication modes, we need to be clear about the identity of the competitors themselves. Only two forms of communication are depicted in Figure 6.1. Two others spring to mind, giving four basic classes of interactive communication:

* random communication by chance meeting of two individuals face-to-face
* travel communication whereby several people meet intentionally at the same location
* long-distance communication by some form of telecommunications or electronic media
* postal communication

Since the first two categories necessitate physical travel on the part of the individuals involved, they may be combined. The latter two comprise interface communication.

The substitution possibilities between travel communication and interface communication largely reduce to answering the following question:

> To what extent can the use of interface communication substitute for face-to-face contact?

A difficulty arising here is to define a single "market" in which these two modes of interaction compete. Many markets are pertinent. For this reason, it makes little sense to talk generally about the substitution possibilities. Meaningful insights require a case-by-case approach within specific sectors or submarkets.

For example, a 1971 survey in Japan revealed that interface communication (telephone and mail) accounted for 84 percent of inter-company business communications. The remaining 16 percent devoted to travel communication was expected to drop to 3.6 percent by 1985. A comparable study by the Direction Générale des Télécommunications (DGT) in France, which also considered intra-company communications, showed that only 8 percent of their communication took the form of meetings requiring travel. Trends in household expenditure patterns show that the average rate of growth in the use of telecommunications in the OECD countries is between 2 to 2 1/2 times higher than the average rate of growth in the use of transportation services. These few observations suggest that the share of travel communication in total communication is declining.

6.3.3 The Comparative Costs Argument

Although it is difficult to construct an aggregate comparative costs model (owing to the heterogeneity of modes), a simplified approach is to compare cost trends in the two dominant modes: road transport and the telephone. Between 1970 and 1980, the gap favouring the telephone widened considerably. A 40 litre tank of petrol in West Germany rose in real cost-equivalent terms from 140 basic telephone charging units to 260 units. Such price comparisons may be of limited significance on their own. But when they are coupled with higher production costs in the motor and railway industries, transportation may be fully recognized as a rising cost sector. By way of contrast, some reduction in telecommunications costs could occur through expansion in network capacities assuming realization of potential scale economies.

6.3.4 The Comparative Energy Consumption Argument

Little data are available to allow the comparison of transport and telecommunications with regard to <u>indirect</u> energy consumption (e.g. for providing infrastructure or maintaining networks). An INSEE study, based on an input-output approach, suggested that the impact of a 100 percent increase in energy costs would be much higher in the transport sector (viz, a 13% price rise on rail, road and waterway services and about 20% on air and sea) than in the telecommunications sector (only about 2%).

Concerning comparative energy consumption, savings of close to 100 percent can be expected by using a teleconference system for a meeting between distant cities instead of travelling to a common city (Katsoulis, 1976). Audioconferencing is cheapest, followed by the videotelephone. In another study of the latter, Dickson

(1973) estimated that an air trip between New York and Los Angeles to attend an 8 hour meeting would consume nine times more energy than a videophone conversation lasting the same time.

Telecommunications do offer attractive alternatives, at least in terms of energy and cost, to the transportation of people and information.

6.3.5 Assessing the Scope for Substitution Growth

Studies in Canada, France and the United States have estimated coefficients of trip substitution by calculating the socio-occupational breakdown of the labour force (up to 1980) and then allocating to each category (or trip purpose) a coefficient for working at home (see Table 6.2). All coefficients are theoretical maxima. The findings of these and other studies are rather similar: they suggest that systematic use of telecommunications would make it possible to replace between 15 to 20 percent of urban trips.

Although business trips represent a relatively small proportion of all intercity trips, they do account for a relatively large share of passenger traffic by air. This mode of transport is the high-growth area, so it is important to assess the scope for the replacement of business trips by teleconferencing. We shall restrict our attention to meetings only since (according to a Bell Canada study) 75 percent of business trips are made for this purpose. Taking an estimate of teleconferencing costs in 1985, Tyler (1978) calculated that 36 percent of all business meetings in the UK could be replaced by remote meetings, corresponding to about 25 percent of business travel. Similar studies in the USA and Canada have suggested that 20 percent of business meetings could be replaced. So the new telecommunications media will be far from negligible as substitutes for intercity business trips.

With the extremely rapid development of facsimile transmission technology, there is greater likelihood of extensive substitution of telecommunications for mail. The basic trends were illustrated earlier (see Figure 6.1 and Table 6.1). Estimates in the seventies were based upon replacement by electronic mail, and ranged from 13 to 40 percent. These forecasts were constrained by the fact that many telephone calls required confirmation in writing. Facsimile services overcome this problem, so much so that the scope for replacement of conventional mail is almost unlimited. As postal services are substantial consumers of transport services, the substitution effect in this context may be considerable.

Table 6.2 Theoretical Substitution Coefficients for Urban Trips

	Trip Purpose						
	Home/ work	Business	Shopping	Social/ leisure	Education	Other	Total
FRANCE (1970)							
Trip Distribution Frequency	50	20	12	12	6		100
Coefficient of Substitution	0.18	0.18	0.20	0	0.35		0.171
Modified Trip Distribution Frequency	41	16.4	9.6	12	3.9		82.9
USA (1969)							
Trip Distribution Frequency	35.2	12.7	17.2	34.9			100
Coefficient of Substitution	0.26		0.60	0			0.228
Modified Trip Distribution	35.4		6.9	34.9			77.2
USA (1975)							
Trip Distribution Frequency	50		22	25	3		100
Coefficient of Substitution	0.24		0.20	0.05	0.25		0.183
Modified Trip Distribution Frequency	38		17.6	23.8	2.3		81.7
CANADA (1973)							
Trip Distribution Frequency	21.6	16.8	15.0	16.6	8.0	22.0	100
Coefficient of Substitution	0.20	0.10	0.05	0.10	0.05	0.10	0.11
Modified Trip Distribution Frequency	17.3	15.1	14.3	14.9	7.6	19.8	89.0

Sources: Adapted from BCEOM-DATAR (1970), Mitchell (1969), Lathey (1975), Wise (1973).

6.3.6 Summing Up

From the (admittedly scant) evidence presented, it is clear that the direct and indirect effects of the development of telecommunications on transport demand are by no means negligible. We have presented illustrative examples, but could have cited many more. On average, these studies suggest that up to 20 percent of journey-to-work trips could be replaced by work at home or at neighborhood work centres. Around 20-25 percent of business travel could be avoided by using telecommunications. Perhaps 15-20 percent of shopping trips could be replaced by teleshopping.

We have also presented arguments based upon comparative economic principles. In terms of unit costs to the consumer, the gap favouring the telephone continues to widen. This may be coupled with steadily rising production costs in the motor and railway industries and the rapidly escalating congestion in air traffic. Savings of almost 100 percent on energy consumption may be achieved by the use of an audio conference system instead of a meeting between distant residents.

But perhaps the clearest evidence of further substitution growth can be gleaned from average per-capita trends in the aggregate demand for travel. Taking Sweden as an illustrative example, Figure 6.2 reveals the mature state of the travel market. The trend line for 1960-75 is considerably flatter than that for the 1950-65 period.

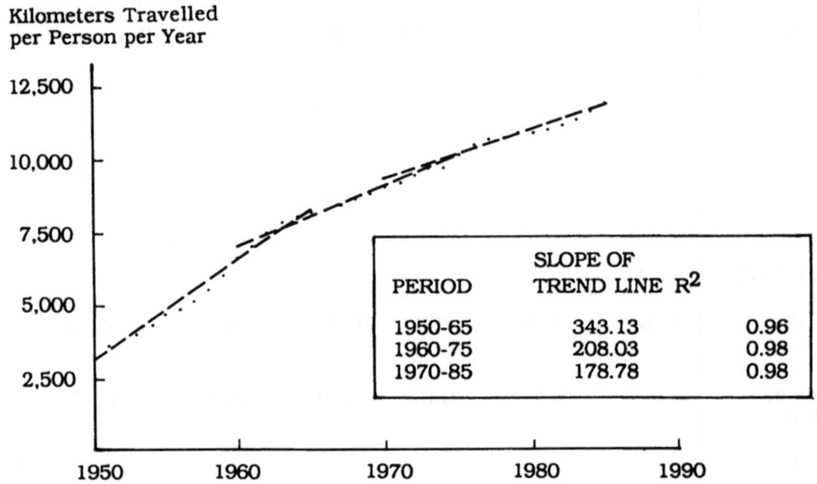

Figure 6.2 Travel Demand per Person in Sweden (Kms per Year)

A further slowdown is evident after 1977. This picture is one of an industry entering a mature phase in its own lifecycle, as it nears its demand capacity (see also Figure 6.3). Although this saturation capacity is not known a priori, it can be calculated using the logistic model described in the Appendix. For Sweden it is about 16,000 km/person/year. Part of its traditional growth in market share has been lost to a competitor. The only serious competitor in the market facilitating human interaction is, of course, interface communication.

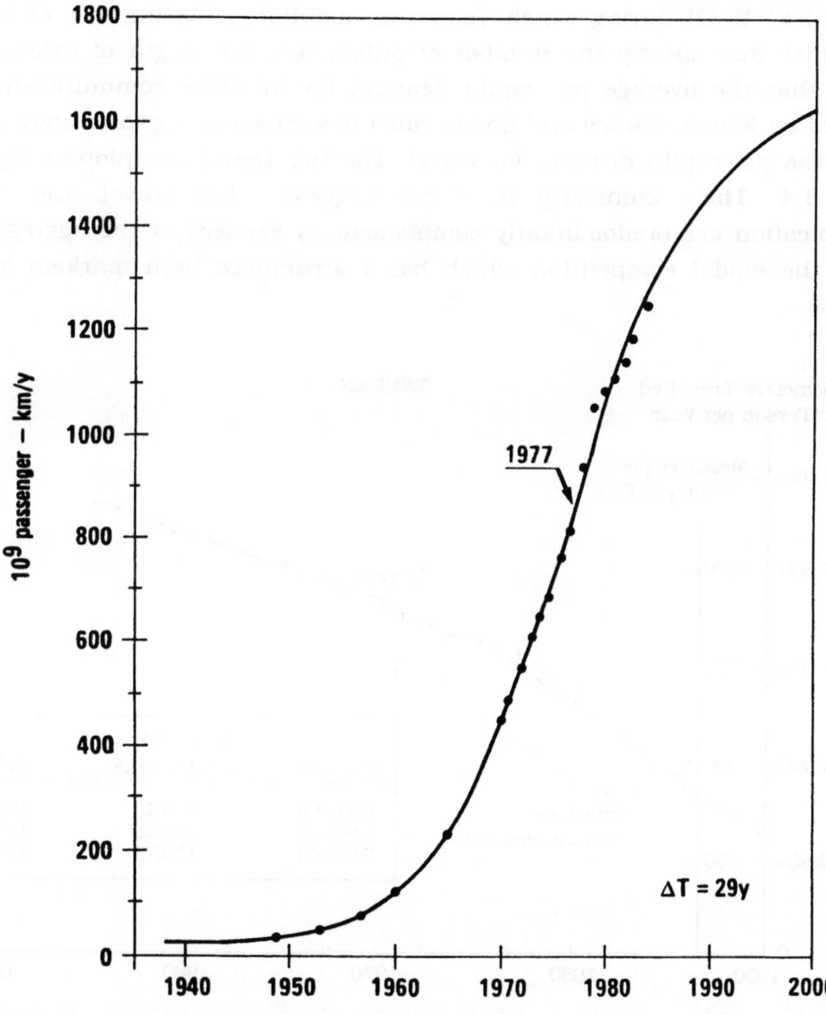

Figure 6.3 World Air Transport, all operations (10^9 passenger-km/year)
Source: Nakicenovic (1986)

6.4 THE CASE AGAINST SUBSTITUTION GROWTH

6.4.1 The Per-Capita Demand Argument

It suits the proponents of substitution abatement to begin their case by referring to precisely the same data presented at the close of the plaintiff's summing up (Figure 6.2). Their opening statement refers to aggregate trends in both travel demand and the use of interface communication. With changes in Televerket's monitoring methods, time-series data available in Sweden for comparative purposes are disturbingly short. But the evidence is nevertheless significant. Over the fifteen year period (1957-1972) during which Televerket monitored the number of telephone calls rather than merely the number of pulses (i.e. the length of calls), we may observe that the average per-capita demand for interface communication (here measured by letters, telexes and phone calls) has grown at approximately the same rate as the per-capita demand for travel. The two trends are plotted together in Figure 6.4. Their similarity in slope suggests that travel and interface communication are predominantly complementary services at the aggregate level, despite the modal competition which has restructured both markets internally over time.

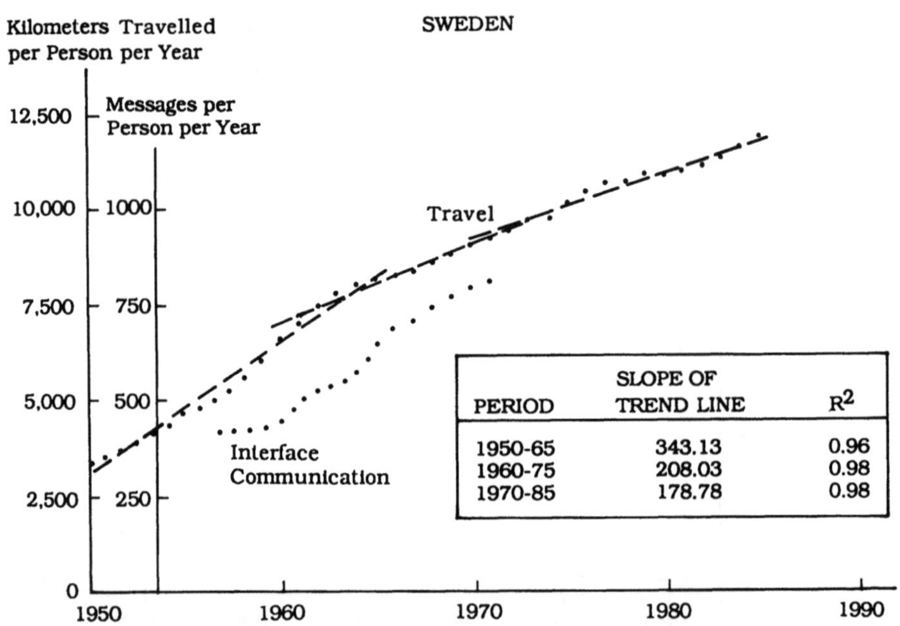

Figure 6.4 Per-Capita Demand for Travel and Interface Communication in Sweden, 1950-1985

Further evidence of the symbiotic relationship between travel and interface communication is displayed in Figure 6.5 The two indexes derived from French data are very similar in nature to the Swedish measures depicted in Figure 6.4. They each reflect the per-capita growth in demand for transport and communication services. The complementary character of these trajectories is clearly evident, spanning almost two centuries of observation.

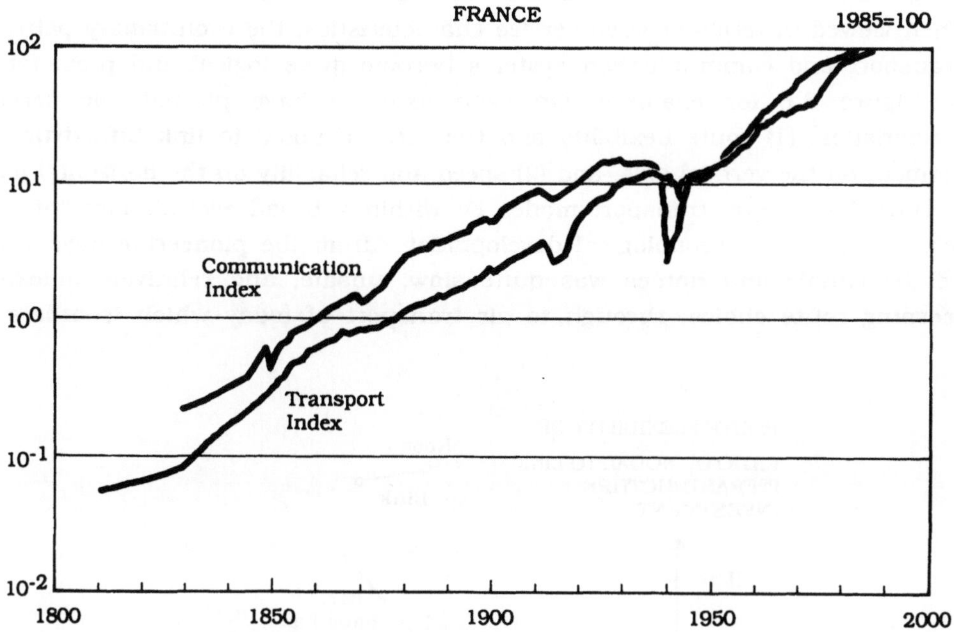

Figure 6.5 Per-Capita Demand for Transport and Communication In France, 1800-1985

6.4.2 The Service Attributes Argument

The driving forces behind these evolutionary trends in modal competition and the aggregate complementarity of transport and communications become more apparent once we conceptualize the demand for each in terms of a bundle of key service attributes. Consider the chronological evolution of transport infrastructure. In addition to the relative cost comparisons which are generally quoted to explain such competitive trends (see, for example, Figure 5.6 in the previous chapter), it is important to understand the underlying trends in service characteristics such as speed, range, punctuality, reliability, route flexibility and value-to-weight ratio. With respect to transport services, a persistent desire to increase each of these characteristics has pervaded the market where new modes have gradually replaced old ones. The advent of air transportation has also heralded a fundamental shift from the physical constraints of surface transport to the relatively unconstrained medium of the atmosphere. This not only guarantees a substantial increase in route flexibility, but it also alters the nature of infrastructure investment: from a high concentration on link infrastructure (such as canal locks, rails and road pavements) to a high concentration on nodal infrastructure (such as airports).

When viewed in terms of such service characteristics, the evolutionary patterns of transport and communication systems become quite logical and predictable. Take Figure 6.6 for example. On each axis we have plotted two service characteristics: (1) route flexibility and the ratio of nodal to link infrastructure investment on the vertical axis, and (2) speed and reliability on the horizontal axis. Note how the various transport modes lie within a broad evolutionary "tunnel" which depicts their chronological development - from the pioneering days when travel by canals and horses was quite slow, unsafe, and relatively inflexible concerning route choice, through to air transport of today which is orders of

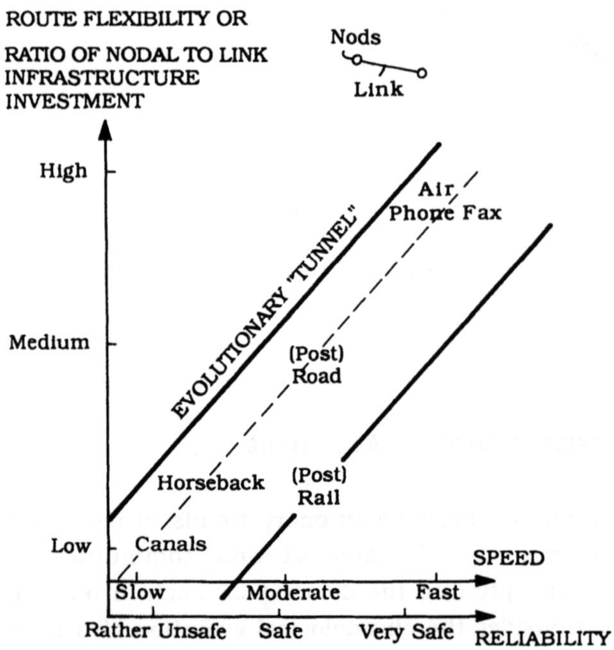

Figure 6.6 Evolutionary Pattern of Transport and Communication Systems in Terms of Service Characteristics

magnitude faster, potentially reliable, very flexible concerning route choice and requiring more investment on nodal rather than link infrastructure.

When the chronological development of communication systems is examined from a similar perspective, an almost identical pattern of changing attributes emerges. The pioneering modes in this case were the messenger on foot or on horseback - clearly demonstrating the complementarity of transport and communication in those early days. The advent of the postal system brought an order of magnitude improvement in terms of reliability and route flexibility, since the organization of the mail was given system-wide (i.e. governmental) support. But the need for cooperation with the transport system remained paramount, since letters cannot deliver themselves. It is tempting to believe that the technology of telecommunications ushered in a new era dictating that the transport-communication nexus would eventually fade. This would, however, be a short-sighted view. The key advantage offered by the (mobile) telephone is very similar to that offered by the air transport system: liberation from reliance on the traditional surface modes of movement. The medium has changed (just as with air transport) in pursuit of faster, more reliable and flexible means of communication offering greater destination choice possibilities.

Figure 6.6 depicts the striking similarity between the evolutionary patterns of transport and of communication systems, when analysed in terms of similar service characteristics. The very fact that both systems' evolution can be measured meaningfully using a common bundle of attributes (changing in a similar direction) confirms that, on aggregate, the two systems are predominantly complementary rather than competitive. Goddard (1973) found that contact patterns of telephone calls and face-to-face meetings (necessitating travel) tend to reinforce one another rather than act as substitutes (Table 6.3).

Table 6.3 Geographical Distribution of Telephone and Face-to-Face Contacts

Destination	% of telephone contacts	% of meetings	Number of respondents
Within Central London	58	64	705
Within Greater London	19	15	
Within Southeast region	8	5	
Within U.K.	12	9	
Overseas	3	7	

Source: Goddard (1973, page 164).

Conferences of short duration over longer distances may be more economical using telecommunications, but the longer the duration and the shorter the distance, the more economical travel becomes (see Table 6.4). In this respect, the scope for substitution between transport and communication is ultimately limited by the importance of face-to-face contacts for "creative" communication, and the extent to which short-versus-long meetings prove fruitful.

Table 6.4 Conference Costs with Origin at Los Angeles

Destination	Duration of conference	Costs	
		Travel	Telecommunica-tions Device
		$	$
San Francisco	2 hours	72.50	53.70
	4 hours	72.50	117.10
Chicago	1 day	370.53	214.94
	2 days	395.53	429.88
New York	2 days	498.23	450.42
	3 days	523.23	675.63

Source: Nilles et al., (1976, page 133).

6.4.3 Further Contradictory Evidence

In section 6.3.3, it was suggested that the comparative cost situation during the seventies swung heavily in favour of the telephone at the expense of road travel. The figures appearing in Table 6.5 (for the years 1970 and 1980) were used to support this argument. With the benefit of the more recent figures included in Table 6.5, such claims are easily refuted. Fuel costs dropped substantially during the early eighties (although all reductions were not always passed on to the consumer). Thus the volatility of oil prices suggests that attempts to draw comparative cost conclusions of the type referred to above may be unreliable, unless based on a much longer time series of statistical observations.

Section 6.3.5 refers to studies which have estimated coefficients of trip substitution for various trip purposes (see Table 6.2). Results suggest that systematic use of telecommunications would make it possible to replace about 15 to 20 percent of all personal urban travel. But the spread of estimates is rather large. Furthermore, each of these studies ignores the additional travel induced by the use of telecommunications, which must be significant if Goddard´s (1973) findings concerning the similarity of contact patterns for telecommunications and travel are to be taken seriously. A Canadian study by Wise (1973) reached the conclusion that more trips would be made by the year 2000 if the induced effects were considered in combination with substitution effects (Table 6.6). Although Wise's estimates assume significant induced effects (such as that social/recreational travel will increase by about 40 percent), they do not consider all the increases in travel encouraged or facilitated by newer communications technology; nor do they include the growing length of trips made per capita each year.

Table 6.5 The Relative Costs of a Telephone Call versus a Tank of Petrol

Year	Average cost of filling a 40 litre tank of petrol	Equivalent number of basic telephone charging units
1970	DM 25	140
1980	DM 55	260
1985	SEK 184	184
1986	SEK 155	155

Source: European Conference of Ministers of Transport, 1982, Swedish Statistical Yearbook, 1988.

Ever since the telegraph was introduced to enhance the safety and efficiency of the railroad, communications have promoted mass travel. Computerized systems are employed widely for mass transit. The mobile telephone releases the businessman from his office, encouraging greater mobility. Facsimile machines are beginning to appear in trucks and aeroplanes, suggesting that a greater share of business may be conducted during travel. The advent of mobile communication systems provides a strong catalytic stimulus towards more travel.

As final testimony to the weaknesses underlying the arguments of the substitution growth advocates, we may refer to a study of the Quebec university system in which it was found that 50 percent of teleconferences were not substitutes for travel, but preparations for a face-to-face meeting, a follow-up to such a meeting, or else an independent use of the system. The popularity of much of the substitution hypothesis stems from its attractiveness to the general media, and possibly from its promulgation by agencies with vested interests (Salomon, 1985).

Table 6.6. Telecommunications: Substitution for and stimulation of urban travel

Relative Index of Trips by Purpose	Trip distribution frequency (1965)	Induced movement %	Substituted movement %	Net change in frequency of travel by purpose %	Relative index of trips by purpose: "electronic future"
From home					
Journey to work	21.6	-	- 20	- 20	17.3
Personal business	16.8	+ 10	- 10	0	16.8
Social, recreational	16.6	+ 40	- 10	+ 30	21.6
Shopping	15.0	-	- 5	- 5	14.4
Education	8.0	+ 20	- 5	+ 15	9.2
From elsewhere	22.0	+ 10	- 10	0	22.0
	100.0 trips				101.3 trips

Source: Wise (1973).

6.4.4 Summing Up

Whereas the case for increased substitution can be argued quite convincingly at the disaggregate level (where specific substitutes for certain types of travel may prevail), the aggregate view suggests that substitution growth is a myth. The overall pattern of travel may be changing, but the level of aggregate demand is still growing. Figures 6.4 and 6.5 illustrate our argument succinctly. The aggregate effect of telecommunications on transport is likely to be a modification of travel patterns rather than any decrease in travel demand.

6.5 REACHING YOUR VERDICT

The evidence presented on both sides is admittedly scant and selective. It is also subject to incessant change. Any juror's view of this debate will therefore be subjected to his or her own prejudices, as well as to an ongoing clash of fact, theory, opinion and personality. Is it really possible to reach a convincing and correct verdict in the face of such a plethora of conflicting perspectives?

There are perhaps two plausible routes by which a reasonable verdict might be reached. The first we shall call the economic criterion. This is encapsulated in the following statement (see Dupuit, 1844): "The ultimate aim of a means of communication must be to reduce not the costs of transport, but the costs of production." This is as true today as it has been in bygone centuries. The establishment of a more economical (i.e. cost-efficient) means of communication alters the sources of supply, so that a simple comparison of the relative costs of transport by different modes (say by canal and by parallel road) is necessarily wrong for the vast majority of products. A similar comparison between the costs of transport and of telecommunications is also necessarily misleading for precisely the same reasons. This economic criterion reveals that substitution possibilities cannot be appraised merely by comparing the relative costs of each alternative. The system-wide implications of each alternative must be taken into account.

The second argument - which we shall label the behavioral criterion - is based on the need to find a means of monitoring the system-wide effects, but at the level of the individual. Travel and interface communication are components of a larger socio-economic system of human behaviour which may be analysed at the aggregate level but must be interpreted at the disaggregate level. A key notion may build upon the classical notion of "space-time prisms" introduced by time-geographers (e.g. Hagerstrand, 1970). In this respect, it is important to understand how the "action space" and "time budget" of an individual or a firm is changing with technological advances. For example, in a typical 24-hour day, it is possible for a hypothetical individual to travel further and to send or receive more messages today than a few years ago. This is an outcome of technological progress (Figures 6.4 and 6.5 confirm that this is actually happening). But what are the time-space bounds to this growing interactivity and can we fit even more opportunities within the same bounds? To answers these questions, we must consider the interface between the barriers of space and the efficient uses of time, which should offer penetrating new perspectives on the complementarity of transport and communication systems.

The reader is left to consider your own verdict, although the author´s bias must be obvious by now.

APPENDIX: THE MATHEMATICS OF MULTIMODAL SUBSTITUTION

Let x_i denote the demand for transport mode *i*. Define a market for transport services as a group *I* in which the various modes $i \in I$ compete to privide a similar type of transport service. Then, at any time *t*, modes *i*'s share of this market is given by

$$f_i = x_i / \sum_{i \in I} x_i \tag{1}$$

so that

$$\sum_{i \in I} f_i = 1 . \tag{2}$$

If the marginal propensity to invest in each mode does not differ appreciably (which would seem to be borne out by the similarity of slopes in Figure 5.3), then we can write down the following system of differential equations (see Peterka, 1977):

$$\frac{d}{dt}\left\{\ln(f_i/f_j)\right\} = \alpha_{ij} \tag{3}$$

where the parameter α_{ij} is representative of qualitative differences (including relative price differences) between modes i and j, and therefore measures the spread over time of the substitution between these two modes. The following symmetric form of the system defined by (2) and (3) may be obtained:

$$\dot{f}_i + f_i \sum_{j \in I} \alpha_{ij} f_j = 0 \quad , \text{ all } i \tag{4}$$

If the coefficients α_{ij} are seen to be reasonably time-invariant, then the solution to (4) takes the following logistic form:

$$f_i(t)/f_j(t) = \exp\left\{\alpha_{ij}(t - t_0) + \beta_{ij}\right\}$$

or

$$\ln\left\{f_i(t)/f_j(t)\right\} = \alpha_{ij}(t - t_0) + \beta_{ij} \tag{5}$$

where t_0 is the time when mode i has captured half of the transport market and α_{ij} is a time positioning parameter.

The above equation means that the logarithm of the ratio of market shares between any pair of modes, *i* and *j*, when plotted as a function of time, should follow a straight line. Such a relationship has been observed to hold for a wide variety of transport and energy markets, including many modal split trends such as those described in Chapter 5 and 6 (see also Fisher and Pry, 1971; Peterka, 1977; Marchetti and Nakicenovic, 1979; Batten and Johansson, 1985).

A detailed economic foundation for this type of simplified analysis may be found in Batten and Johansson (1989). It is in any case clear that if *A(t)* defines the matrix of intermediate deliveries between all sectors and y_j the final demand for mode *j*, then

$$\frac{f_i(t)}{f_j(t)} = \frac{[I - A(t)]_i^{-1}\, y_i(t)}{[I - A(t)]_j^{-1}\, y_j(t)} \tag{6}$$

Letting $C_i(t) = [I - A(t)]_i^{-1}$, we see that

$$ln\left\{\frac{C_i(t)\, y_i(t)}{C_j(t)\, y_j(t)}\right\} = \alpha_{ij}(t - t_0) + \beta_{ij} \quad . \tag{7}$$

Such a simplified model is straightforward to test and has strong appeal as a means of furnishing preliminary projections of changes in the demand for transportation services by mode and location.

REFERENCES

Batten, D. F. and Johansson, B. (1985), Industrial Dynamics of the Building Sector: Product Cycles, Substitution and Trade Specialization, in *Economic Faces of the Building Sector*, (eds) Snickars, F., Johansson, B. and Lakshmanan, T. R. Document D20:1985, Swedish Council for Building Research, Stockholm, 191-215.

Batten, D. F. and Johansson, B. (1989), Dynamics of Product Substitution, in *Advances in Spatial Theory and Dynamics*, (eds) Å. E. Andersson, D. F. Batten, B. Johansson and P. Nijkamp, North-Holland, Amsterdam, 23-44.

BCEOM-DATAR (1970), Etude de la Substitution Transports-Télécommunications, Rapport definitif, April.

Dickson, E. N. and Bowers, R. (1973) The Videotelephone: a New Era in Telecommunications, National Foundation, Cornell University, Ithaca, N.Y.

Dupuit, J. (1844) De la Mesure de l'Utilité des Travaux Publics, *Annales des Ponts et Chaussées*, 8:84-110.

European Conference of Ministers of Transport (1983) Transport and Telecommunications, ECMT, Economic Research Centre, Paris.

Fisher, J. C. and Pry, R. F. (1971), A Simple Substitution Model of Technological Change, *Technological Forecasting and Social Change*, 3:75-88.

Goddard, J. B. (1973) Office Linkages and Location: A Study of Communications and Spatial Patterns in Central London, *Progress in Planning: 1*, 1-185.

Hägerstrand, T. (1970) What about people in regional science? *Papers of the Regional Science Association*: 24, 7-21.

Katsoulis, M. (1976) Travel-Telecommunication Substitution; its Potential for Energy Conservation in Canada, Bell Canada.

Lathey, C. E. (1975) Telecommunications Substitutability for Travel: an Energy Conservation Potential, Office of Telecommunications, Washington D C, 75-58.

Marchetti, C. (1985) Swings, cycles and the global economy, *New Scientist*, 2 May: 12-15.

Marchetti, C., and Nakicenovic, N (1979), The Dynamics of Energy Systems and the Logistic Substitution Model, Research Report RR-79-13, International Institute for Applied Systems Analysis, Laxenburg, Austria.

Mitchell, M. (1969) On the Substitutability of Telecommunications for Urban Motor Vehicle Travel, National Academy of Engineering.

Nakicenovic, N. (1986) The automotive road to technological change: diffusion of the automobile as a technological substitution process, *Technological Forecasting and Social Change*, 29:309-340.

Peterka, V. (1977), Macrodynamics of Technological Change: Market Penetration by New Technologies, Research Report RR-77-22, International Institute for Applied Systems Analysis, Laxenburg, Austria.

Nilles, J. M., Carlson, F. R., Gray, P. and Hanneman, G. J, (1976) *The Telecommunication - Transportation Tradeoff: Options for Tomorrow*, Wiley, New York.

Salomon, I. (1985) Telecommunications and Travel: Substitution or Modified Mobility, *Journal of Transport Economics and Policy* 19:219-235.

Tyler, M. (1978) Implications for transport, in *Impacts of Telecommunications on Planning and Transport*, Department of the Environment and Transport, Report 24, 129-167.

Wise, A. (1973) Impact of Electronic Communication on Metropolitan Form in the City, in *Canada Prospects*, (ed). L. Axworthy, Butterworths.

Chapter 7

USING TECHNOLOGY TO IMPROVE TRANSPORTATION SERVICES

William L. Garrison
University of California at Berkeley, U.S.A.

7.1 THE TRANSPORTATION TECHNOLOGY LIFE CYCLE

The life cycle concept has long been used by technologists and managers of research and development, and it is increasingly used by managers of firms and policy analysts (Ayres and Steger, 1985). It applies to products, and it uses a biological language. A product is conceived as an idea and birthed as a prototype. With subsequent refinement, it begins to be adopted by markets. Eventually, a rather standardized product saturates the market, senescence is boded when sales are mainly replacement ones or when competing products begin to nibble away markets.

7.1.1 Transportation Systems

Application of the life cycle concept to transportation appears straightforward. Indeed, transportation products are often used to illustrate the concept. There was the railroad steam engine development cycle and its replacement by diesel or electric propulsion. The Model-T Ford ran a life cycle until abruptly replaced by competitive products. The highway system in the U.S. is widely regarded as mature, standard designs have saturated the market.

But in spite of numerous examples, the application of the life cycle concept is not neat. For one thing, it is the transportation system product that is at issue, and not just this part or that part of the system. The Model-T ran its life cycle, but the automobile-highway system has many more decades left in its cycle. There are also complications when the relations between technologies and systems' structures are considered.

To aid in extending to a systems frame, transportation can be usefully viewed as produced by a triad of hard and soft technologies:

* There are fixed facilities, such as air and maritime ports and railway and highway routes, and the soft technologies associated with them, such as the operating knowhow held by port authorities.

* Equipment is involved. There are auto and auto manufacturing organizations, air frame manufacturers, etc.

* And there is an operations side. Hard technologies, such as LORAN C for navigation, and soft technologies, such as liner operational and navigation knowhows, are involved. The household's use of an automobile is an operations matter.

To ask about the product we have to ask how these system parts, working together, are related to technology development and to markets.

Bringing in markets also requires a stretch of imagination. Information is readily available on the parts of systems. One can say something about the market for tankships or for automobiles. But consideration of the market for personal accessibility or the resources made available by freight transportation runs to vagueness because the question of what transportation does for society has hardly been explored, as is often the case with complex systems and deep social and economic questions.

7.1.2 Getting Started

How does the life cycle apply to transportation systems? The first phase in the cycle is that of conception and birthing of a prototype. System design and application in a market niche are the central features of system beginnings. The Stenvenson-Pease (1820s) design of the Stockton and Darlington Railroad to serve the Auckland coal fields, McLean's (1950s) design for New Jersey-Texas container shipping, and Fulton's (1800s) Hudson River steamboat come to mind. In each case, designs were system scoped. They involved a guideway and other fixed facilities, equipment, and uses/operations.

The birthing of a system is usually associated with equipment development--Fulton's steamship and Stevenson's locomotive for example. The Model-T Ford and the Douglas Corporation's Model-3 (DC-3) are also cited. But closer examination underscores the importance of system design in market niches. For one thing, one can almost always find workable equipment available prior to system birthing. Steamships, locomotives, containers, and automobiles were available before successful system designs were found, sometimes available for decades. System innovations required more than equipment. They incorporated fixed facility, equipment, and operations hard and soft technologies in successful designs in market niches.

Reference to the air and automobile systems introduces another aspect of system birthing: Sometimes designs emerge from the workings of somewhat independent actors. Adding to development of the DC-3 for the air transportation system, actors involved with airports, airways, navigation aids, airline firms, government and users all contributed to the innovation process. Highway builders, users, and other actors played their roles in the development of the auto-highway system. In addition, the development of these systems occurred in multiple market niches. The auto was a rich man's toy in many cities and soon a work-horse on farms, for example.

So while convenient for short hand description, the tieing of an innovation to equipment development slights the system design process. This point will re-emerge in the last part of this discussion when suggestions for technology development are discussed.

7.1.3 Growth and Development

Once the idea of the system is demonstrated as a prototype, improvements are rapid. The design itself may be modified, as was the case when the Liverpool and Manchester, London and Birmingham, and Baltimore and Ohio railroads eliminated many of the tramway features and modified the common carrier aspects of the Stockton and Darlington. But by-and-large the system design changes little, and improvements are made at the level of the systems parts.

Although a limited snapshot, Figures 7.1 and 7.2, showing information on Atlantic liners, illustrate the pattern of improvements. The passenger and priority freight liner system dates from about 1838 when Brunel introduced the Great Western accompanied by shipbuilding, dock, and organizational developments. As materials became available, ship construction shifted from wood to iron to steel. The engine-propulsion system shifted from paddle wheels to multiple screws; simple steam engines added expansion cylinders, and then gave way to turbines. Velocities increased with more efficient powering, and, as markets grew, liners were increased in size to achieve scale economies. The technologies provided for service improvements of a factor of two or greater.

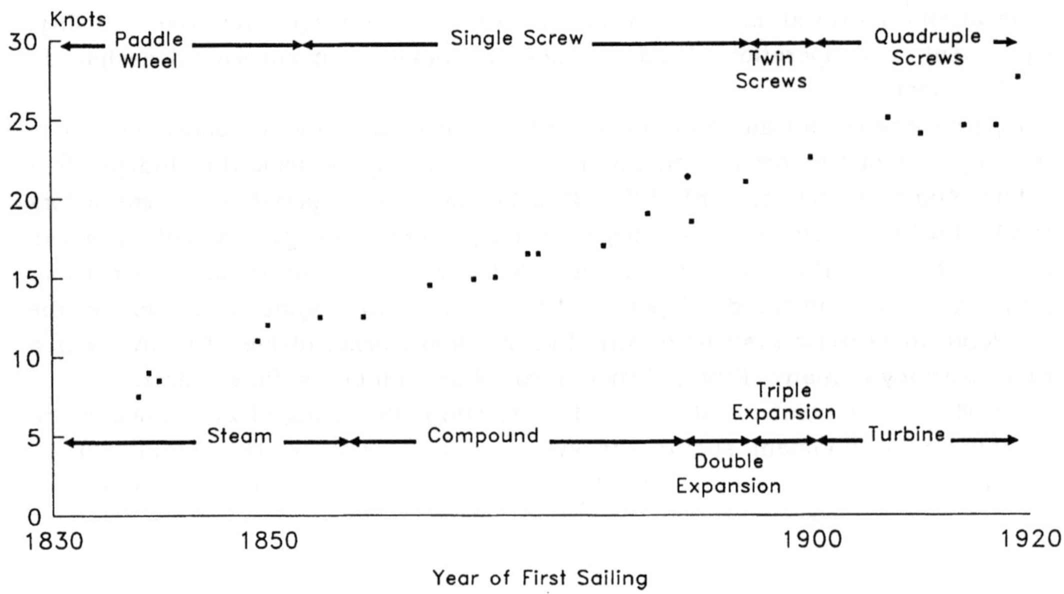

Figure 7.1 Evolution of the Atlantic Liner, 1838-1914: Speed.
Source: Data from Zimmermann (1923).

During this same period, the other parts of the system also improved. Insurance arrangements were modified. In the search for safety, bulkheads were improved (but not enough to prevent the Titanic disaster), and the Plimsoll Line was introduced. Port, navigation, and other improvements continued.

The pattern, partly illustrated by the liner exemple, is that of some revision of the system design early on, but mainly the pattern is that of hard and soft technological improvements specific to parts of systems. Taking the system design as given, technology is developed and introduced to improve system parts. The pay-offs may be lowered costs or greater reliability. There may be service quality improvements.

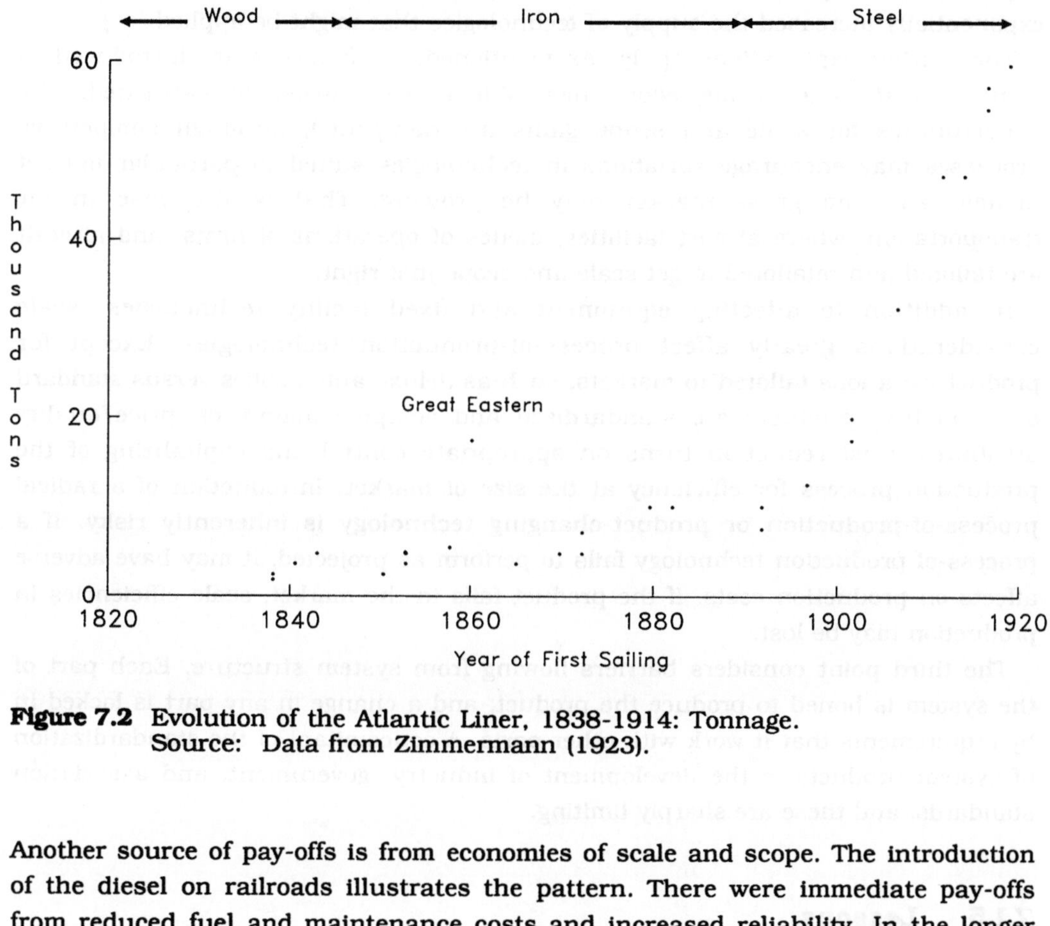

Figure 7.2 Evolution of the Atlantic Liner, 1838-1914: Tonnage.
Source: Data from Zimmermann (1923).

Another source of pay-offs is from economies of scale and scope. The introduction of the diesel on railroads illustrates the pattern. There were immediate pay-offs from reduced fuel and maintenance costs and increased reliability. In the longer run, the diesel allowed longer run-throughs increasing the areal scope of service improvements, and diesels operating in multiple units permitted increased train sizes and weights and achieved economies of scale on dense haul routes.

7.1.4 Maturity

Systems eventually mature by saturating their markets and standardizing their technologies. Standardized services are available everywhere, and growth or retraction of service responds to gross market size. Much of the fates of automobile and transit services in the United States in recent decades is as simple as that.

The technologies become rather frozen in part because the higher pay-off technologies are introducted early-on and diminishing returns set in. That explanation is partial, because work by scientists and technologists has exponentially increased the supply of technologies that might be applied.

Three other explanations apply. As mentioned, technologies are introduced to capture scale and scope economies. When the market is saturated, the opportunities for scale and scope gains are dampened, although competitive processes may encourage variations in technologies suited to particular market niches, and the gross market may be growing. That is the case in air transportation, where airport facilities, modes of operations of firms, and aircraft are tailored and retailored to get scale and scope just right.

In addition to affecting equipment and fixed facility technologies, scale considerations greatly affect process-of-production technologies. Except for product variations tailored to markets, such as deluxe automobiles versus standard ones, mature products are standardized and compete mainly on price/quality attributes. Cost reduction turns on appropriate control and capitalizing of the production process for efficiency at the size of market. Introduction of a radical process-of-production or product-changing technology is inherently risky. If a process-of-production technology fails to perform as projected, it may have adverse affects on production costs. If the product fails in the market, scale efficiencies in production may be lost.

The third point considers barriers flowing from system structure. Each part of the system is honed to produce the product, and a change in any part is locked in by requirements that it work with other parts. A concomitant of the standardization of system products is the development of industry, government, and association standards, and these are sharply limiting.

7.1.5 Lessons

The consideration of the life cycle highlights quite different roles for technology as systems move along the life cycle. Early-on, the task is that of embodying technologies in system designs. That is an integrative effort that uses building blocks whereever they can be found; its a try-out-a-design-in-a-market-niche activity. It treats system parts as clay and it molds them (Sahal, 1981).

Once a successful prototype is found, technologies are introduced to modify designs and, especially, to improve system parts. Gains in system performance are rapid as the system begins to be deployed and scale and scope economies are achieved. But at "best", standardized designs are found and as the market trends toward saturation, a system is more-and-more locked-in by its structure and by standards and regulations preserving that structure. The best things to do have been done, and most scale and scope economies have been found. More-and-more, windows for technology applications are limited.

Finally, the system matures as a standardized design fills the market. Technology is restrained to low risk applications conforming to system structure. It tends to be focused on product cost and performance attributes and/or differentiating a standardized product to market segments.

7.2 WHAT IS TECHNOLOGY DOING FOR TODAY'S TRANSPORTATION SYSTEMS?

Against the backdrop of lessons from consideration of transportation life cycles, what is the situation in today's transporation systems? How is technology contributing to improvements? What windows for technology applications are available? What should be expected of technology? The life cycle concept is general, specifics are needed.

We shall deal with these questions in three steps. First, today's situation will be critiqued through reference to an expert, consensus study of technology developments. Second, current and expected contributions to improvements will be explored using two modal cases, auto and air transportation. Finally, brief comments will be made on other modes.

Large questions such as those to be explored never answer easily, and there is sufficient diversity among the modes that answers that fit all modes well fit none exactly. Furthermore, a convincing argument is difficult because it yields a counter-intuitive result, and one that is not flattering to industry managers and technologists: Today's work is not inducing gains comparable to those of the past, and there is good evidence that expectations about technology should be modest unless improved strategies for development and implementation are adopted.

7.2.1 The Situation

In 1982 the National Research Council in the U.S. included a chapter on transportation in its Outlook for Science and Technology: The Next Five Years (National Research Council, 1982). The short time frame of the projections and near-concurrence of the time period with today suggest that the chapter provides expert judgement of current developments. The developments summarize:

1. Microprocessors used for automobile engine control extended to highway traffic control for flow improvements. Route optimization and off-board vehicle control may evolve later.
2. Higher speed helicopters for several hundred mile service.

3. Supersonic aircraft.
4. Vertical/short take-off and landing aircraft for relief of airport congestion.
5. Improvements in the air traffic control system; improved in-cockpit devices.
6. Advanced design, large scale bulk carriers for the Great Lakes.
7. New rail equipment for intermodal service.
8. Technologies to reduce stress on rail.
9. Improved intermodal system management techniques.
10. Improved dewatering and use of fluids other than water in slurry pipelines.
11. Energy efficiency improvements in vehicles from reduced weight and aerodynamic, powertrain, and engine improvements; lighter, stronger materials.
12. Development of alternative fuels.
13. Reduced atmospheric and noise emissions.

Leaving aside the points that a different study committee might have produced a somewhat different list, and that some of the listed developments have not yet appeared, what signals is the list sending?

Equipment development dominates the list, and technologies embodied in products rather than processes-of-production also dominate. We hesitate to draw a conclusion from this observation. Equipment is largely produced by competitive firms, and such firms seek visibility for their work; equipment is much discussed. Also, there is a tendency to associate technology with physical products, and especially equipment rather than fixed facilities or operations.

Leaving the equipment emphasis aside, three factors to which technology is responding may be identified. First, there is work to improve service in existing markets by specializing equipment to markets, especially aircraft. A second factor is the availability of new technologies, microprocessors and materials. Finally, there is response to changes in the prices of inputs, operations costs, and the costs of output externalities. Real or anticipated increased energy prices press many of the developments; there is response to congestion problems and pollutant and noise emissions.

Perhaps the appropriate summary statement is this: There is use of technology to manage problems and to tailor services to markets. Problems include changes in factor prices, congestion, and undesired externalities.

Two additional questions will now be treated. First, how do the technologies relate to stage in life cycle? Second, what pay-offs should be expected from use of these technologies.

7.2.2 Technology and Life Cycle

The discussion of the life cycle pointed out that systems birthed as designs incorporate fixed facilities, operations, and equipment in market niches. None of today's technology developments appear to have that potential. Intermodal, supersonic aircraft, and helicopter developments, may have that potential if embodied in new system designs.

Once birthed, there are rapid improvements in technologies for the parts of systems, improvements of a factor of two or more, as the development of the Atlantic liner illustrated. As technologies are standardized and as systems increase their markets, economies of scale in production and operations are critical, and technologies are sought that aid achieving economies of scale. None of today's technology developments are sharply pulled by these "best, standardized" or "economy of scale gaining" purposes; none run the risk of running counter to the purposes.

As markets near saturation or become saturated, the manager's problem is that of tailoring services to market segments to the extent that is possible with standardized product and process-of-production technologies. Managers also seek to lower input factor prices and to adjust to changes in their environments in order to remain competitive. Problems must be managed. Most of the technology developments noted respond to these requirements.

This partial conclusion is reached: most of today's technology development is characteristic of mature systems. The conclusion is somewhat arguable because it involves interpretation and classification of developments. There is also a time-will-tell question, especially for intermodal and aircraft developments.

The conclusion that today's technology developments are characteristic of mature systems was termed a partial conclusion because there is another part to the conclusion. It is that the consequences of technology development will not be great. The "consequences will not be great statement" is a comparative one. Early in the life cycle the birthing of new systems and subsequent rapid technology improvements improve services or lower costs by a factor of two or more, as mentioned. Though limited in impact by this comparison, today's improvements are worth pursuing, of course.

7.2.3 Consequences of Today's Technology Developments

At this point, the statements just made about the consequences of today's technology developments are assertions. To flesh them out, the situation in the auto-highway system will be examined in some detail, short remarks will be made on the air system, and summary sentences will be given on the other modes. Space prevents treating all the modes, and it limits what can be said about the modes taken as cases.

Considering the automobile-highway system, its history and its technological history in particular are well known. In the U.S., the road part of the system began to be improved to accommodate the automobile during the 1910s. Cities developed protocols for the delivery of roads and streets. Arterial and local access road classification and design standards emerged by the 1920s. Urban institutional and financing schemes were developed.

Rural roads began to be improved using designs, surfaces, and structures suited to the automobile. Roads that were at best Telford-macadam designs were improved as the federal and state governments assumed responsibility for the state systems, at first, seven percent of rural mileage. Counties or other local governments began to improve other rural roads.

Building from proposals during the 1930s for regional defence and toll road highways, the interstate system was implemented in 1956. Its mileage is largely rural, however, urban extensions of the interstate and other freeway facilities were constructed in the urban areas.

Steam powered automobiles (and busses) were tried out beginning early in the 1800s, and steam, electric, and Otto cycle engine vehicles began to find markets during the first decade of the 1900s. In the American market, Henry Ford's Model-T was introduced in the 1910s. It set the pattern for a mass market vehicle (Otto cycle engine, naphtha fueled, four wheels, engine in front, alloy steel) produced using standardized parts and mass production methods. Traffic control technologies aiding the use of the system also emerged in the early 1900s, and they were refined as users developed the knowhow for system use.

At first the automobile was a good weather, rich man's toy. In the 1910s, the Model-T began to replace the horse and wagon in rural areas, and urban users began to organize work, social, and other trips using the automobile. By that time, suburbanization, outlying shopping centers, and other features of the city reflecting automobile dependence began to emerge.

Although the timing differs from nation to nation, the pattern of the emergence of system parts holds among the developed nations. It's a well-known story, but not one that delves deeply into the stage in life cycle of the system, its use of technology, or its performance. Stage in life cycle can be deciphered by examining the evolution of system parts. The highway system is shown in Figure 7.3. Needed was a product suitable for automobile use, and hard surface pavements are one attribute of such a product, although that measure does not trap highway quality improvements, e.g., wider lanes, as development proceeded. Figure 7.4 provides a fragment of information on the deployment of high capacity, high quality facilities.

Although facilities continue to be built, freeways have saturated their market in Los Angeles.

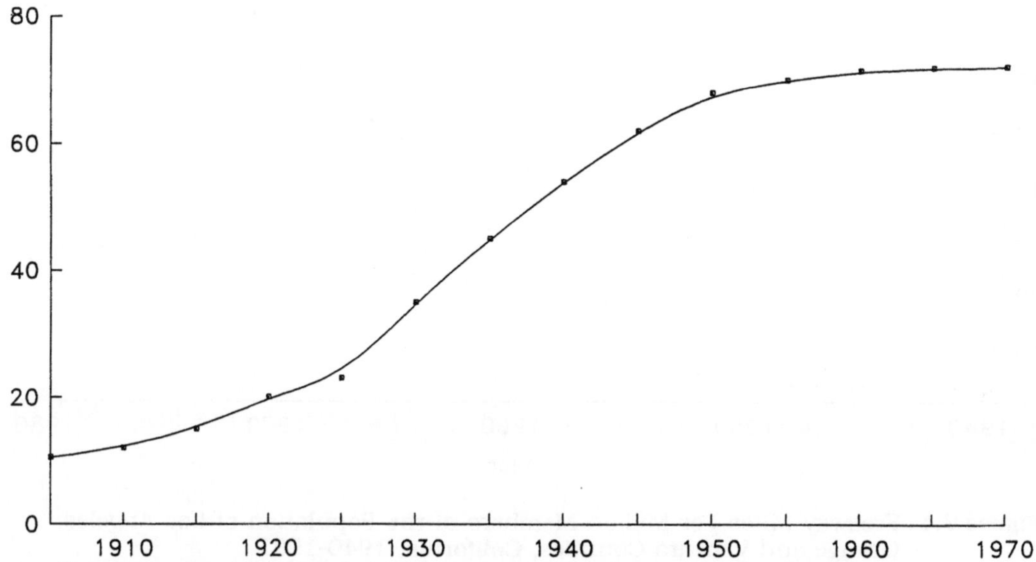

Figure 7.3 Linear Feet of Paved Road Per Capita: 1905-1970.
Source: Data from U.S. Department of Commerce (1977).

During the seven decades of the production of mass-produced vehicles, the automobile has penetrated its U.S. market quite deeply (Figure 7.5). On average, an automobile is available to almost everyone with a driver's licence, although there are variations among households correlated with age of household members and income. Low income and/or older households own fewer than average vehicles.

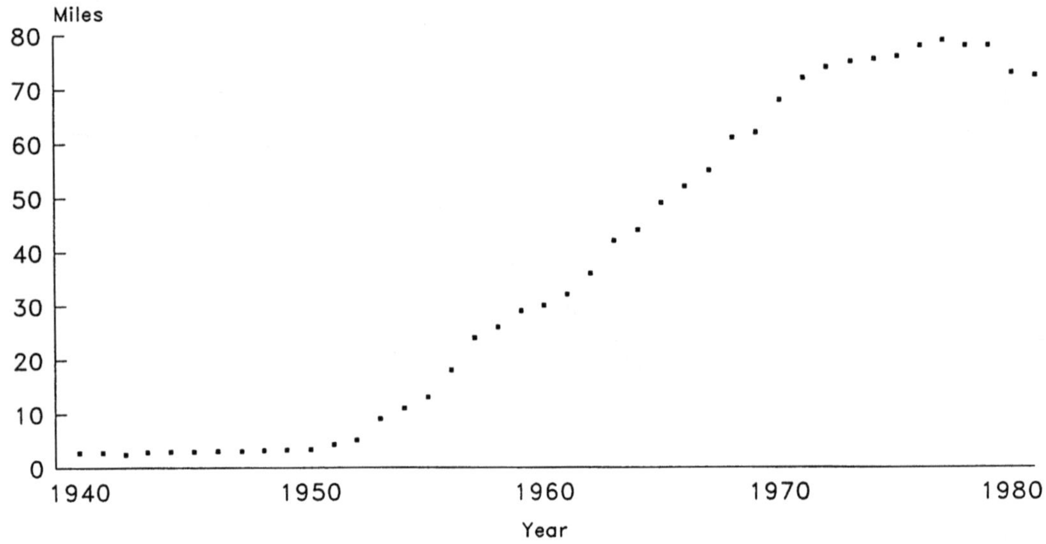

Figure 7.4 Freeway Miles Per Million Members of the Population of Los Angeles, Orange and Ventura Counties, California, 1940-1981.
Source: Data supplied by the Highway Engineering Department of the Automobile Club of Southern California.

The trend in the use of the system is partly suggested by the availability of vehicles. It is also suggested by the average miles per year vehicles are driven. That is now about 11,000 miles in the U.S., having increased gradually from about 8.6 thousand miles in 1936 (U.S. Department of Transportation, 1987). The use of the system is growing mainly because more-and-more persons elect to use the system and not so much because individual vehicles are used more-and-more. The trend in vehicle miles of travel per capita illustrates how use is tending towards saturation (Figure 7.6).

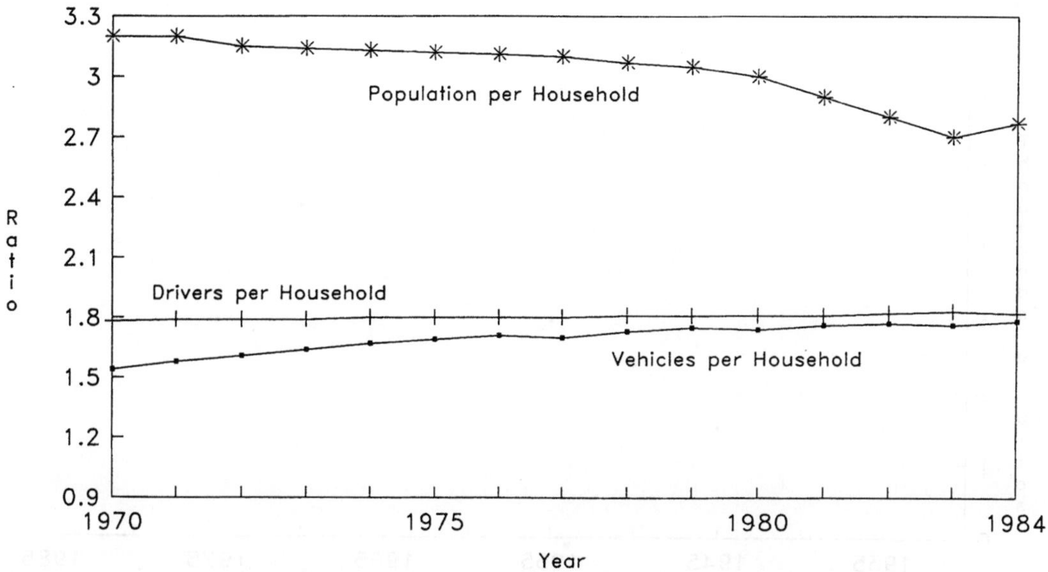

Figure 7.5 U.S. Population, Drivers and Vehicles Per Household, 1970-1984. Source: Data from U.S. Department of Energy (1987).

Market saturation is one characteristic of mature products. Other characteristics include decreasing returns from new technologies and decreased performance. The discussion will now begin to review these.

The life cycle-tied pattern of technology development has been treated. William Abernathy's study of automobile manufacturing adds another dimension to the general pattern (Abernathy, 1976; see also Sahal, 1981). He pointed out that innovations are embodied in products early-on and in processes-of-production as products are standardized. He noted that major automobile hardware developments occurred prior to 1940, and although process-of-production developments began with the transfer line in 1910, they accelerated later than hardware developments and continued longer, indeed to today.

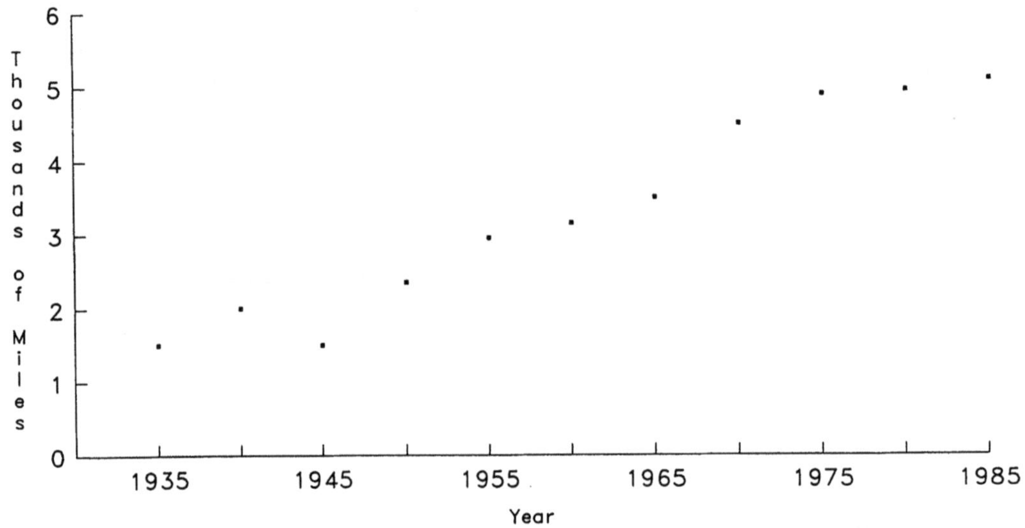

Figure 7.6 Passenger Vehicle Annual Miles of Travel Per Capita, 1936-1980, Charted at Five Year Intervals.
Source: Miles of travel data from U.S. Department of Transportation (1987).

An examination of the road part of the system shows the same pattern. In road building, the first problem was to get the product right (materials, construction techniques, etc) and then to standardize it (Seely, 1984). That effort gave sharp returns into the 1920's and 30's, when freeway designs emerged (Gifford, 1985). Beginning in the 1920s, processes tied to scale of production must have become an important source of improvements. Clearly, the production process was refined (Johnson, 1978; U.S. Department of Transportation, 1977).

Ideally, we would like to be able to tie life cycle technology development to the performance of the automobile-highway system. Costs and quality are critical performance measures. Is the real cost of the service decreasing; is the quality improving; or is there some favorable mix of these? Productivity trends provide related measures. Is a given quantity of output being produced using reduced inputs or is output increasing from a given level of inputs?

A number of studies of productivity trends are available (Hooper, 1987), but, with the exception cited below, these do not match the system structure as discussed in this manuscript. Also, by focusing on inputs and outputs, productivity studies take a "black box" view of changes within activities. To better connect modal performance to stage in life cycle, systems, and technology, data will be used that show trends that can be tied to technology.

The California Department of Transportation has calculated a highway construction price index (CHCPI) beginning in 1946, and that index is compared to the consumer price index (CPI) in Figure 7.7. Briefly, construction costs lagged the CPI until the middle 1970s, and it has escalated above the CPI since. A study of construction productivity trends for U.S. Federal-aid highways found total factor productivity increasing until about 1965 and decreasing subsequently (Kane, 1978).

The productivity and cost trends can be tied to technology, at least in a rough way. Moavenzadeh refers to playing-out of mechanization and management advances initiated the 1920s (Moavenzadeh, 1985). He also refers to highway construction program expansion and the achievement of scale economies. Other sources highlight similar trends (Public Works Historical Society, 1986; Johnson, 1978). With the build-up of post World War programs, improvements flowed from scale effects. After the early 1970s when the pace of construction began to founder, scale-based improvements began to be lost.

The situation in automobile production is shown in Figure 7.8. The factory sales value series shows sharp decreases in the real cost of the product until about 1930, reflecting product and process-of-production technology improvements during the first two decades of the life of the automobile product. The 1930-1950 period was one in which the product was redefined by technology initiatives and pulled by market response. (The former are discussed by Moritz and Seaman, 1981.) In a sense, the 1910-1930 product was largely a Ford Model-T derivative, and today's product is the derivative of a product revision during the 1930s. During the 30's, the engine and passenger compartments were moved forward relative to the axles, and modern brakes, automatic transmissions, suspension, radios and heaters, and metal bodies began to characterize the standardized product.

Consumer expenditure data are available for the post War period, and these suggest a mild continuation of the more-car-per-car trend (Motor Vehicle Manufacturers Association, 1987).

Nakicenovic points out that the substitution of the automobile for horse drawn vehicles was virtually complete by 1930, and he refers to the post 1930 period as one of the diffusion of individual transportation. He then analyzes the diffusion of quality improving technologies, such as disk brakes (Nakicenovic, 1986). Without rejecting the importance of quality improving technologies, we think of the post 1930 period as one in which development was pulled by the innovation of new, rather that substitute, uses for the automobile. The important innovations were in the use/operations part of the system as the population began to do things not practicable using horse drawn vehicles. Physical realizations of new uses were the auto suburbs and shopping centers dating from the 1920s. New uses and improved vehicles pushed developments in the road part of the system (Gifford, 1984).

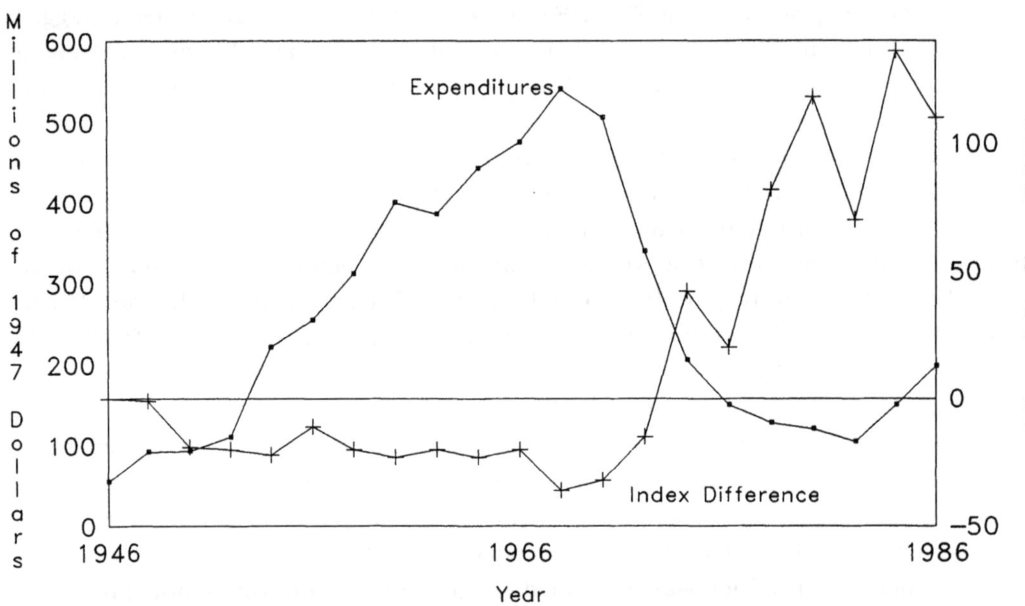

Figure 7.7 California Highway Expenditures in 1947 Dollars and the Difference Between the California Highway Construction Price Index (CHCPI, 1947 = 100) and the Consumer Price Index (CPI, 1947 = 100), 1946 to 1986.
Source: Data from Jones (1988).

With respect to the costs of system use, real costs of vehicle operations have remained stable since 1950 (Figure 7.9). Isolated data points for earlier decades suggest sharp decreases in costs earlier in the product life cycle (Ogilby, 1924; National Automobile Chamber of Commerce, 1921).

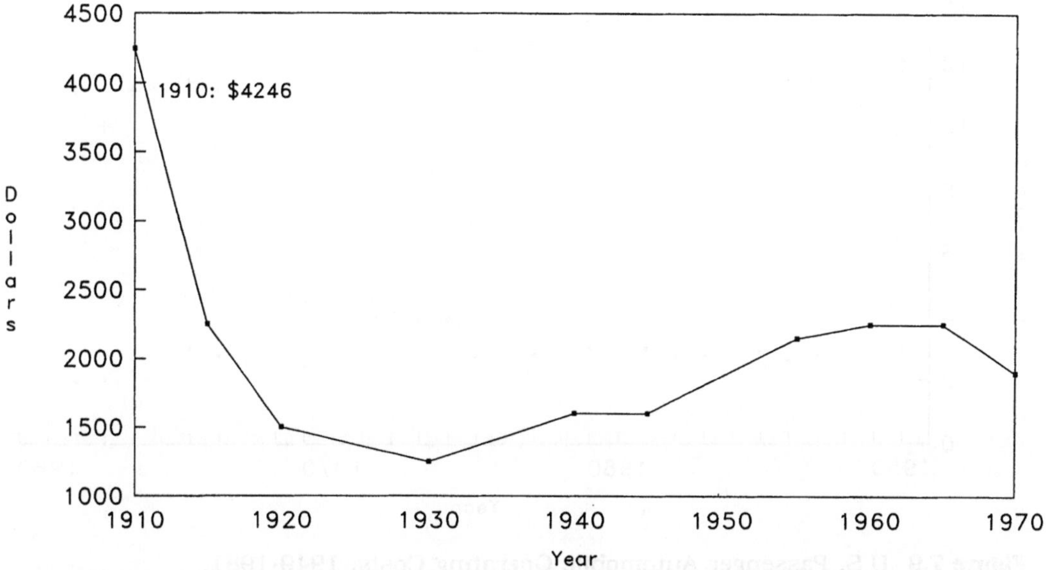

Figure 7.8 Average U.S. Factory Sales Value of Domestic Passenger Vehicles in Constant 1967 Dollars: 1910-1970.
Source: Data from Motor Vehicle Manufacturer's Association.

The purpose of the discussion above was to position the automobile-highway system in its life cycle and to inquire about the past and present contributions of technology to product improvements. The summary statement is that technology made consequential contributions in the past, especially in lowering real costs (1910-1930) and/or improving quality (beginning in the 1930s). In comparison, it is not doing much now.

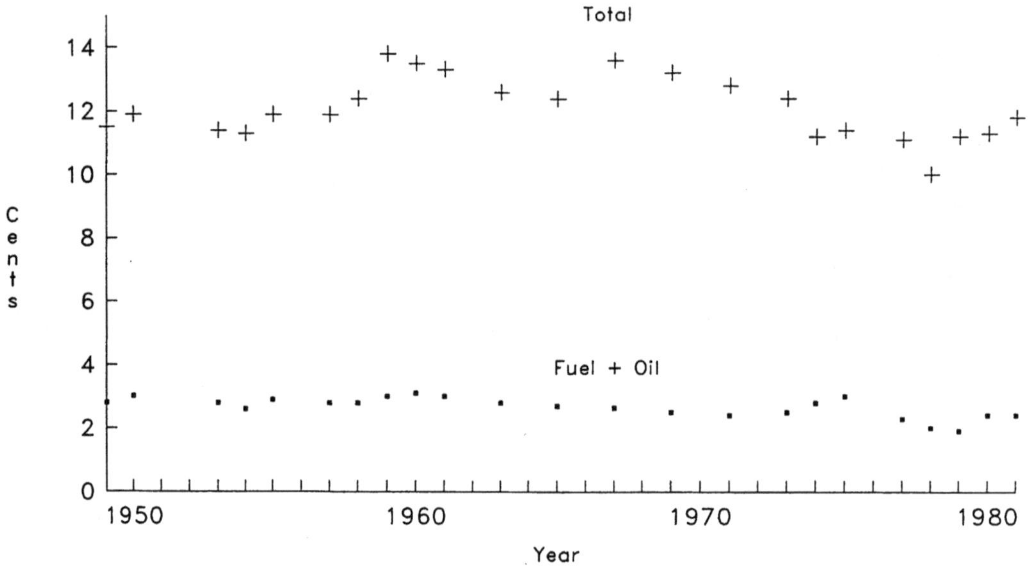

Figure 7.9 U.S. Passenger Automobile Operating Costs, 1949-1981, in Constant 1967 Cents Per Mile.
Source: Data from the Motor Vehicle Manufacturers Association.

That statement can be disputed in two ways. The analytics are not crisp. The uses of index numbers and comparison of products to the CPI beg considerations beyond those made; the comparison of technology introduction with product costs was a general one. In addition, quality improvements in products have not been fully considered. In particular, both the automobile and the highway have improved in quality during recent decades, though both were building on freeway and automobile designs introduced in the 1930s. Even so, the broad conclusion certainly holds.

7.2.4 The Air Transportation Case.

To present the air transportation situation in a brief fashion, Figure 7.10 displays the trend in carriers' expenses per domestic revenue passenger mile.

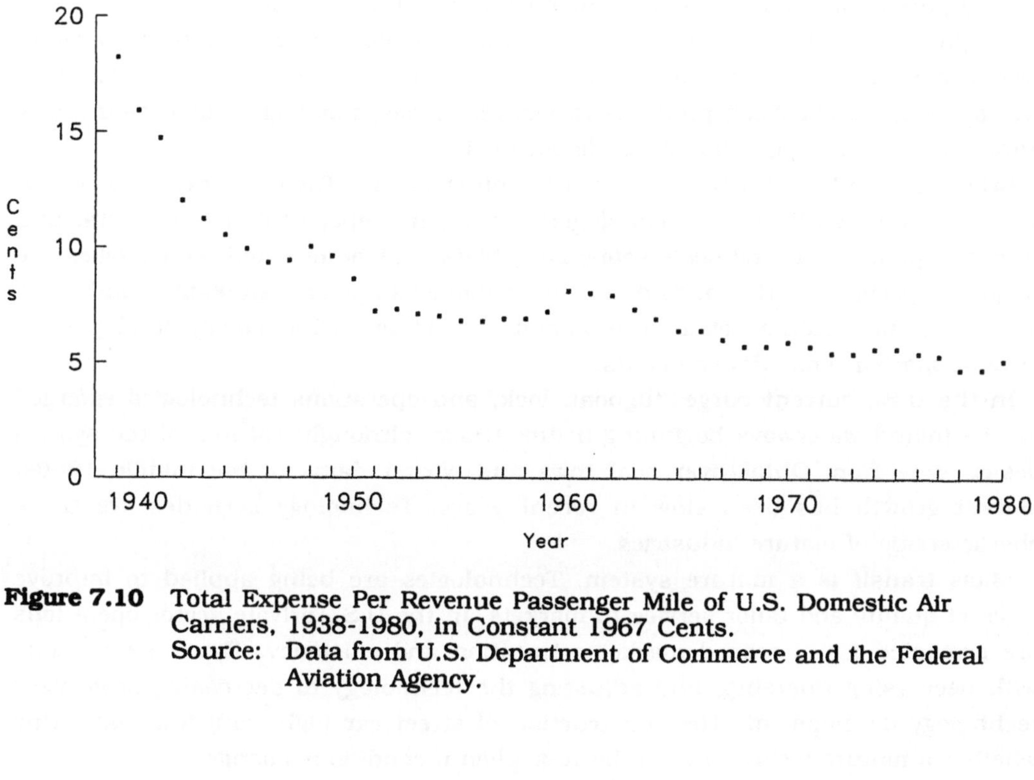

Figure 7.10 Total Expense Per Revenue Passenger Mile of U.S. Domestic Air Carriers, 1938-1980, in Constant 1967 Cents.
Source: Data from U.S. Department of Commerce and the Federal Aviation Agency.

The introduction of DC-3 type aircraft and subsequent improvement and scale-up of that technology yielded a J-shaped trend from the beginnings of the commercial system in the 1930s to the 1960s. Equipment improvements interacted with technology improvements in operations, air traffic control, airport, navigation, and other parts of the system design that emerged in the 1930s.

The introduction of jet aircraft in the late 1950s occasioned a brief period of higher costs, but the cost trend has continued more-or-less along the J-shaped trend initiated in the 1930s. The introduction of jet aircraft did sharply improve the quality of the product. Flight stage lengths were increased, velocity increased, and high altitude, smoother flight was obtained. Safety has steadily improved.

The introduction of jet aircraft also occasioned a partial system redesign process. Airports were enlarged, air traffic and navigation protocols were changed, and firms adjusted operations. Today, there is a search for technologies to flesh out these changes, reduce energy use, and grasp scale economies. Industry deregulation has accelerated that search, and the advantages claimed for deregulation are mainly a result of that acceleration. Aircraft designs, for example, are scaled and tailored to market niches. Hub and spoke airport and operations designs strive to accommodate scale economies from high volume passenger throughputs in airports and along routes (Morrison and Winston, 1986).

In spite of its relative youth and rapid market expansion, the air transportation system has the earmarks of a maturing system, and the chief roles for technology are tailoring standardized products to market niches, reacting to increased factor prices, and supporting obtaining scale economies.

Brief statements will now be made about other modes. The container ship system design is well established. Technologies for larger ships, container carriers, and cranes support achieving scale economies. Market niche oriented technologies are sought, especially on the landside of the container shipment. Tankships and other bulk ships have been specialized to markets and to scale. Technology developments have supported those developments.

In the U.S., current barge, tugboat, lock, and operations technologies emerged on the inland waterways beginning in the 1930s, although features of the system design date from Ohio River coal tows and wicket dams of the middle 1800s. Market growth has been slow in recent years. Technology activities are those characteristic of mature industries.

Mass transit is a mature system. Technologies are being applied to improve product quality and tailor service to markets. In the U.S., private sector operations are promoted for increased market orientation and efficiency. There are markets with decreasing ridership, and adjusting the technology to decreasing scale begs technology development. The construction of street car (light rail) lines is asking whether a mature technology can be re-applied if conditions change.

Shifts in quantities of commodity flows and market locations press for pipeline technologies appropriate for diverse scales. The U.S. railroads (and pipelines) differ from the other modes in that fixed facilities, equipment, and operations are under corporate control. Although technology mainly finds markets in system parts, there is some opportunity to jointly tailor system parts to market niches. The development of unit trains, using routes and operations specially developed for them, is an example of technologies providing for the tailoring of system products to markets.

The truck-highway system is introducing technologies which may be found in each of the modes. Pipelines, for example, have adopted sensor, computer network flow analysis and control, and materials developments. A major thrust in railroad operations is the introduction of new technology control systems, as mentioned.

Information systems enable improved passenger and commodity flow control in all the modes.

7.3 OPPORTUNITIES

This previous section began with a list of current technology developments produced by a National Research Council study. It then explored a question posed by that list: What are these developments contributing to transportation improvements? It goes without saying that the technologies are making contributions, otherwise they would not be developed. Technologies are improving safety and reliability, providing for product differentiation among markets, lowering costs or holding back cost increases driven by higher factor prices, and improving service quality generally. That there are improvements is not the issue; the issue is compared to what and whether pathways for major system improvements are being opened.

To provide for comparisons, products were positioned in their life cycles and fragmentary data on technology contributions to product improvements over product life cycles were examined. Conclusions are straightforward: 1) The opportunity for consequential improvements is created when technologies are embodied in full system designs. 2) Opportunities are captured as the "best", standard or standardized, or predominant technology is created and as technologies enable capturing economies of scale. 3) Once the product is mature and standardized, technology based improvements are real, but small compared to the previous contributions to improvements.

Today's transportation systems are mature. Developments consistent with maturity are one option for the future. Given the high level of "push" from today's technology developments and "pull" by system problems, opportunities are great. Even so, expectations of improvements from technology should be modest compared to improvements obtained in the past.

Consider the automobile, for example. Much is made of current changes in the automobile manufacturing industry. Daniel Jones, one of the principals in the Massachusetts Institute of Technology study of the future of the automobile (Altshuler et al, 1984), concludes that the industry is "dematuring" (Jones, 1986). In contrast, we interpret the changes as those of an internationalizing mature industry. The industry is coming to terms with the competitive implications of regional differences in factor prices and in the ways hard and soft process-of-production technologies have been honed. It continues the search begun in the 1930s for product quality improvements and the differentiation of a standardized product to market segments.

At any rate, the automobile is one part of a larger system. The standardization of the system puts sharp limits on changes in the automobile; it must fit the system. The issue is system maturity.

7.3.1 New Products to Obsolete the Old

As a biological concept, the life cycle is accepted as inevitable and irreversible. The long sweep of history does not deny that, for product after product has run the life cycle path. However, history also reports that old products are obsoleted by new ones, and the use of technology to create new services that obsolete the old provides one option for the future.

In the 1800s, railroads and associated feeder roads and waterways obsoleted the waterway-tramway-road system of the 1700s, the container liner system of the 1960s obsoleted a good part of break-of-bulk shipping, and the DC-3 system of the 1930s obsoleted previous air services. Considering that the designs of most transportation systems are quite old, many dating from the turn of the century or before, efforts to obsolete old services provide an attractive option. Although modified as time has past, fundamental features of systems reflect the states of precursor technologies at the time of their births, the nature of markets at the time, and factor prices of the times.

The railroad, for example, was created using a suitable gauge for the times; a few light-weight cars were trained behind head-end power. Seeking mainly scale economies within the given design, today, heavy and high-center-of-gravity cars are incorporated in long trains. The results are difficult train handling and track maintenance problems; heavy, strong equipment is required to handle the compression and tension forces in long, heavy trains. In addition to equipment maintenance costs, energy use, damage to shipments, derailments, and car sorting problems follow from the present-day version of an old design.

The electric telegraph was not available when the railroad was innovated, and its adoption by railroads in the middle 1800s did sharply affect organizational structure and management (Chandler and Salsbury, 1965). Today's communications technologies are also sharply affecting railroads. Even so, railroads system designs are little modified.

A close look at almost any system reveals the heavy hand of the predominant design and technology as determined by past conditions. Put simply, if the system were to be innovated today, it would be different.

7.3.2 Breaking the Tyranny of the Life Cycle

Abruptly obsoleting a system and its product is a disruptive way to circumvent the limitations product maturity imposes on opportunities for technology. It obsoletes investments, institutions, and individuals, as well as products; it may trigger recessions (Santini, 1985). Continuing changes in system designs would yield a more desirable path for change. Continuing changes in designs would avoid the locking-in of standardized technologies for systems and their parts, would track product improvements on markets and changes in resource prices, and would permit full use of technologies, including those that do not fit a locked-in design. Development would be less linear from old designs and more interactive with the environment (Pacey, 1983).

A main factor in locking products into standard ones is the economy of scale achieved from producing a given product. Another factor is economy of scope--by standardizing, systems can operate as networks and achieve efficiencies by integrating operations. Although in the interest of efficiency, network-imposed standardization produces the "one product fits all situations and none well" character of transportation service.

Ayres and Steger propose that emerging computer, robot, and artificial intelligence technologies hold the promise to eliminate the inevitability of the life cycle (Ayres and Steger, 1985). The technologies promise efficient small scale product. Perhaps there is also promise in these technologies for the elimination of standardization.

7.3.3 Barriers

A discussion of barriers is obligatory in an analysis of this type, and with respect to circumventing the constraints imposed by the life cycle, the rule that technological change must be incremental may appear to be an overriding barrier. It is not.

A variety of reasons are given for the incremental rule, running mainly to the point that a technology has to fit into a given situation. That, of course is no more than a comment on the life cycle behavior of systems. The standardization of the product and the technologies producing it define incremental markets for technologies.

Considering incremental change in the broad sweep of transportation history, even those changes that had revolutionary consequences were incremental. Everything incorporated by McLean in the design of container shipping, for example, was incremental. The container was an incremental development from the truck body, the Ideal-X; the first container ship was a T-2 tankship incrementally changed by adding a spar deck derivative of decks developed for shipping aircraft by sea. The soft technologies for operations and management built incrementally from truck and maritime technologies (Kendall, 1986).

Indeed, in container system development, many of the changes were so slight that, at best, they should be termed minor incremental changes. That is the case elsewhere. The DC-3 itself differed from precursor designs hardly at all. It differed from the Boeing 247 mainly in size, for example. What was different was that a variety of aircraft developments where fused in an aircraft design at the right scale. Then, existing navigation, operations, management, and air traffic control developments came together into a system design.

So the rule that technological change must be incremental does not constrain new system designs. Designs portending consequential improvements may involve some incrementalism or may not. The design is the new technology.

Transportation systems and their institutions are large and complex, yet observations of the development of successful technologies leads to the rule that technologies are developed most successfully by small, independent entrepreneurs (e.g., U.S. Congress, 1980). The image of the innovator in the garage holds. If technology development is to occur, that is how it must begin.

An interacting observation is that systems are so large and complex that single entrepreneurs, or even large firms, cannot do much (National Research Council, 1984). This calls for massive efforts coordinated by governments or some other large powerful organization. The U.S. man-on-the-moon or Defense Department style is necessary.

These observations become irrelevant when the pattern of revolutionary change or system innovation by design is noted, but design introduces its own barrier. Design calls for a market niche. The single innovator cannot put the system together in a design in a garage and see if it works from the viewpoints of markets and the technology itself. Stevenson and Pease had the Auckland coal fields of Northeast England for their market niche, McLean had the Texas-New Jersey empty container market niche, Fulton had Albany-New York on the Hudson, and the air system had the American Airlines sleeper-service market niche.

In the face of highly standardized systems and services, the barriers to exploring design options in market niches are very real. Constraints on imagination are a major barrier. The recent mature or near-mature behaviors of systems provide the experience experts and publics use for thinking about technology; the problems that technology might manage and the things technology might do are judged on recent experiences.

Also, the services provided by existing systems limit ideas on what systems might do. Returning to system design innovations, they occurred in market niches where substitution for existing services provided a competitive opening, but they, and subsequent follow-on developments, explored and found markets for new services. Stevenson and Pease built the Stockton and Darlington to move coal, they were surprised to find that passengers also wanted to move. The early automobile was a substitute for the buggy and a rich man's toy. It took a while to learn how the automobile might be used for more. Matson thought that three containerships moving about 400 containers each would saturate the Pacific market (Stindt, 1983). It took a decade to learn what container shipping could do.

This inability to understand markets is the major reason for defense style, step-by-step moves toward a known goal; technology development is in appropriate for transportation.

We judge the limitations on imagination imposed by recent experience and the need for market niche testing as the critical barriers to major technological improvements in transportation.

7.3.4 Suggestions

A major suggestion emerges directly from what has just been stated. It is that innovation and technology development be posed as a system design and trial in market niche activity. An operative word is system, and system design refers to manipulating fixed facility, equipment, and operations. Compared to activities conforming to opportunities provided by the current stage in life cycles of systems, a design in a niche activity holds potential for order of magnitude improvements in systems.

With that guide in mind, this discussion will now turn to suggestions. To align the discussion with current interests, familiar developments will be treated-applications of electronics to highway transportation and the development of small energy - efficient passenger vehicles and integral trains. The main objective in making suggestions is not to argue for priority designs; it is to illustrate attributes of promising designs.

Smart Highways; Smart Vehicles. For two decades or so, vehicle manufacturers and traffic engineers have been accelerating the uses of electronic technologies--automating engine controls and improving traffic sensitive traffic lights, for example. Using that experience and rapidly evolving technologies, there is interest today in combining vehicle navigation, vehicle identification and position sensing, and communication and computer technologies to improve traffic control and traffic flow in congested urban areas (Strobel, 1983). These automated highway system (AHS) developments are responding to technological opportunities and congestion problems. Such ideas are elaborated in succeeding chapters.

Also during recent decades, there has been rapid development of automated guided vehicles (AGV) or mobile robots for use in factory or warehouse environments (Warnecke, 1983). The technology is similar to the AHS technology.

AGVs utilize optical line following, wire following, or signpost location aids, they communicate with and are controlled by computers.

Although the technologies are similar, there are two important differences between the AHS and AGV developments:

> The AGV is embedded in a system design, and is part of a rearrangement of production processes. The AHS is the adding of electronics to an existing system.
>
> The AHS is oriented to managing problems-congestion and drivers or truck dispatchers limitations on finding efficient routes. While the AGV does manage problems of material positioning and movement, it is part of a larger effort to explore and grasp opportunities for new modes of production.

The critical difference between the AHS and AGV activities is that the AGV activity has greater similarity to past activities that have seeded revolutionary change in transport.

How might the automated highway effort take on more of the system design flair of work with the AGV? The niche available in and in the vicinity of container ports would enable a system design scoped effort. The entering wedges for automation are the efficiencies to be obtained in managing queueing, requirements for timely positioning of containers, and requirements for precise information. Port authorities treat the port area as a design problem, so there is the opportunity to design operations, fixed facilities, and equipment as whole.

Efforts in several ports would provide diverse experiences and seed automation in many areas. The port experience might extend from these seeds into urban goods distribution and management and regional commodity shipments.

Freight Transportation: The railroads in the U.S. are hard-pressed by competition from other modes, and more-and-more they are specializing in bulk freight and trailer- or container-on-flat-car movements. For these and other reasons, individual railroads and the Association of American Railroads (AAR) sponsor active research, development, and technology implementation programs.

There is also the development of an integral train concept (Association of American Railways, 1984). In the past, locomotives and cars were designed as the market demanded. With concentrated markets for hauling, such as for coal or containers, the idea has surfaced of designing trains for specific markets, rather than designing locomotives and cars separately. Features discussed for these trains include application of power to axles throughout the train (distributed power), cars permanently coupled, and designs specially tailored to markets--loading and unloading capabilities, for example.

Although design scopes to equipment and operations, its consideration of fixed facilities is limited. There are, of course, track construction and maintenance implications for the routes on which trains might be operated, and some terminal design implications. In most respects, the integral train development is consistent with mature product improvements--it tailors the product to markets.

With full extension to fixed facilities, however, the integral train concept could have a system design character. One system design might involve automated car movements with near constant dispatch and receiving. Building from existing clearances, cars might have a gauge of, say, ten feet and be capable of moving 400 tons, in the case of bulk movements. Removing the necessity to manage the compression and tension forces from car training, requirements for car strength would be reduced compared to conventional cars. On-board diesel-electric power might be used at first (motors and steering on each wheel), with off-board electric power a later possibility. For low value bulk commodities, high throughput could be obtained at low velocity.

Numerous market niches are available. Increasingly, for example, building material aggregates are moved greater distances to urban markets. There are niches in those markets. Coal has been mentioned. Striving for just-in-time logistics in manufacturing and warehousing yields niches.

Reconfiguring the Auto-Truck-Highway System. Building from conditions in the early days of the system, highways accommodate both large trucks and automobiles, but neither as satisfactorily as separate automobile and truck systems would. The accommodation of trucks requires low grades, wide lanes, and strong pavements and bridge structures. But even modern highways limit desired truck size and weights. The presence of trucks interferes with automobile operations and creates safety problems as 1- to 2-ton automobiles mix with 40-ton trucks. Generally, and especially in urban areas, congestion is a problem.

Suppose an urban route was reconfigured to serve only personal vehicles. There is discussion of double-decking urban freeways, and these might be car-only facilities. Removing the conflicts between trucks and automobiles, smaller vehicles become a possibility and lanes might be narrowed. Dealing with the stability problem of small vehicles by leaning and, perhaps, using electronic lane-keeping aids, a 500 pound, high velocity, energy efficient vehicle, such as the Lean Machine developed by General Motors, might become practicable (Figure 7.11). Using limited space for parking and reduced lane requirements, the capacity of the roadways might be more than doubled.

At the same time, truck-only highways might be developed from unused rail right-of-way or carved from existing road spaces. Special interregional routes or facilities in the vicinity of ports or industrial districts might evolve. Such facilities might take on AHS features specialized to them, perhaps, features evolving from AGV systems.

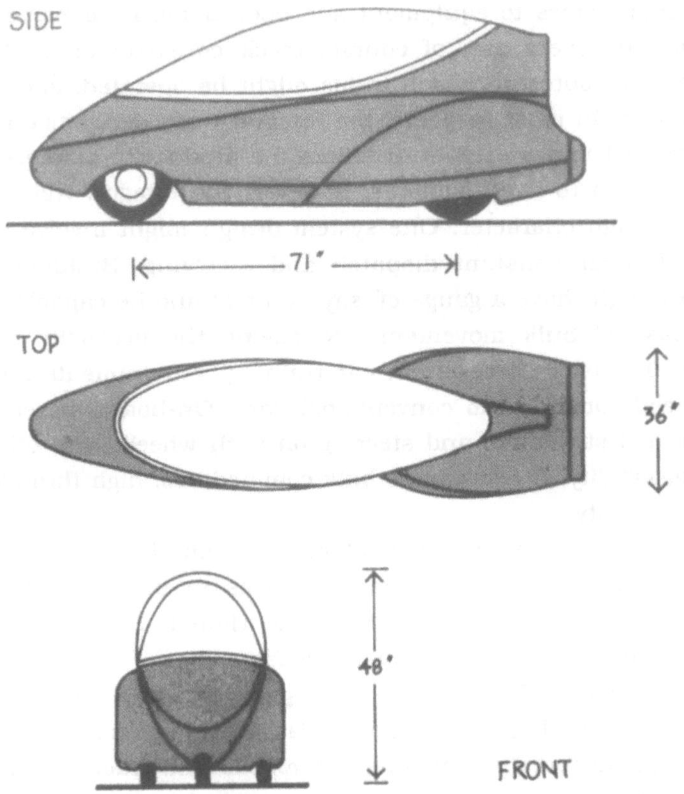

Figure 7.11 A Possible Commuter Car: The General Motors Lean Machine

7.3.5 What the Suggestions Say

Presented only in outline, the suggestions have a grasped-out-of-thin-air character. However, they were selected because they illustrate several attributes that history says are important to major technology advances. In the background provided by previous discussion is this paper, the attributes identify easily.

First, the suggestions are conservative in that they build from existing and emerging technologies. They are also conservative in that they depend, at most, only on incremental improvements of technology.

Second, all of the examples are system design oriented. The first order building blocks are the major parts of systems--fixed facilities, such as guideways and terminals; equipment; and operations. Although not emphasized, technologies that crosscut or fit parts are also building blocks. Although also not emphasized, the designs involve both hard and soft technologies. An automated port, for example, would demand development of appropriate soft operations technologies.

Third, each example was market niche oriented. Each was oriented to a niche, or many similar niches, where the three major or first order parts of the design might be manipulated.

Fourth, each promised efficiency or product quality improvements as a substitute for existing activity. Short term pay-offs could motivate development.

Fifth, each might open options, as opposed to technology developments that strive to use technology to "mine-out" gains from further development of a given, standardized system design. Further exploration of the suggestion for redesigning the auto-truck-highway system will illustrate important aspects of this point, the open options point. The passenger vehicle described could be considered a commuter car - an inexpensive, high velocity, reduced-energy-and-space-using vehicle for commuting. The idea for a system incorporating that vehicle can be expanded to systems serving a variety of social purposes. A neighborhood car-road system could be imagined in which low velocity, environmentally benign, easy to drive, inexpensive and fuel efficient vehicles serve for school, shopping, and social trips.

Extending, systems can be thought of yielding many vehicles for every garage and suitable guideways and modes of operations. Road capacity should not be a barrier, for specialized vehicles would make efficient use of existing space.

There are two points to be made now. One is that a design in a niche would open-up social considerations of options--options that are now hidden because of the lack of stereotypes of what might be done. It is in this sense that a new design "seeds". It is also in this sense that, as a sociotechnical system, transportation innovations are social innovations. Given an idea of what might be, interaction with markets drives design decisions.

The second point is that the design in a niche should be consistent with emerging social and economic trends and with resource situations. In addition to taking advantage of modern manufacturing methods and materials, a commuter car could be very energy efficient and, relative to existing vehicles, benign with respect to noise and air pollution. Shortages of well-positioned urban land would be eased with a high velocity, low congestion-creating and space-using vehicle.

The inexpensive neighborhood system would be consistent with the aging of populations and demands for widened mobility for persons regardless of driving skills or monetary resources.

Increased specialization in all things is a deep running social trend--jobs, recreation, and education (Moore, 1963). The birthing of specialized systems is consistent with that trend. Proposals for small cars are not new, and there have been efforts to market them. The present proposal is different, for it is a system proposal, a design involving guideways, vehicles, and operations.

7.4 CLOSURE

This discussion began with an examination of the transportation product life cycle. The examination of the windows for technology available as a product moves along its life cycle and the improvements induced by those technologies revealed that:

1. System innovations require only off-the-shelf technologies, although they may involve some development; their key features are the system scope of the designs and the testing of designs in market niches. They seed major opportunities for technology-induced improvements.
2. Once the system is innovated, there are major opportunities for technology to improve system parts, the processes-of-producing those parts, and the services provided. As products become standardized, there are opportunities to develop technologies to support achieving economies of scale and scope. These technologies permit achieving factor decreases in costs and/or improvements in product quality.
3. As a system begins to saturate its market, standardize its product, and mine-out scale and scope economies and other sources of efficiencies, the market for technologies becomes more and more limited. However, increases in the availability of technologies, as is occurring rapidly today, and the escalation of problems as mature systems attempt to react to changed conditions, opens opportunities for cost effective activities. In a sense, the window for opportunities grows smaller as the pressure to use technology grows.

With this pattern in mind, the discussion turned to suggestions for improving the applications of technology to transportation. Suggestions or examples of applications were given. These emphasized testing system designs in market niches, especially designs that might seed new developments.

REFERENCES

Abernathy, W. J. (1978). *The Productivity Dilemma*. Baltimore, MD: Johns Hopkins University Press.

Altshuler, A. et al, (1984). *The Future of the Automobile*. Cambridge, MA: MIT Press.

Association of American Railroads (1984). *High-productivity Integral Train Project*. Washington, D.C. US. The association.

Ayres, R. U. and Steger, W. W. (1985). "Rejuvenating the Life Cycle Concept--And Possibly the Economy", *Journal of Business Strategy* Summer: 66-76.

Chandler, A. D. and Salsbury, S. (1965). "The Railroads: Innovators in Modern Business Administration", in Bruce Mazlish, ed., *The Railroad and the Space Program*. Cambridge, MA: MIT Press.

Garrison, W. L. and Taff, S. (1988). *Assessment of Technologies for Highway Transportation Operations*. University of California, Berkeley, CA: Institute of Transportation Studies.

Garrison, W. L. (1987a). *The Performance of Urban Highway Transportation*. University of California, Berkeley, CA: Institute of Transportation Studies.

Garrison, W. L. (1987b). "What History Says About Transportation Technology", *Technology Options for Highway Transportation Operations*. University of California and California Department of Transportation, Berkeley, CA: Institute of Transportation Studies.

Gifford, J. H. (1985). "The Innovation of the Interstate Highway System", *Transportation Research* 18A: 319-332.

Hooper, P. G. (1987). "Productivity Change in Transport: A Survey", *Transport Reviews* 7: 341-367.

Jones, D. T. (1986). "The Internationalization of the Automobile Industry", *Entrepreneurship in a "Mature Industry"*. University of Michigan, Ann Arbor, MI: Center for Japanese Studies: 113-131.

Jones, D. W. (1988). *Compilation of Data on California's Highways*. Berkeley, CA: Privately published.

Johnson, L. R. (1978). *Memphis to Bristol: A History of the Tennessee Road Builders' Association*. Nashville, TN: The association.

Kane, A. R. (1978). *Cost Trends and Productivity Advances in the Federal Aid Highway Construction Industry: 1956-1975*. Washington, D.C., US: The George Washington University. Ph.D. thesis.

Kendall, L. C. (1986). *The Business of Shipping*. Centreville, MD: Cornell Maritime Press.

Moavenzadeh, F. (1985). "Research Needs in Transportation Facilities", *Transportation Research* 19A: 497-510.

Moore, W. F. (1963), *Social Change.* New York: Prentice-Hall.

Moritz, M. and Seaman, B. (1981). *Going for Broke: The Chrysler Story.* Garden City, NY: Doubleday & Co., Inc.

Morrison, S. and Winston, C. (1986). *The Economic Effects of Airline Deregulation.* Washington, D.C.: The Brookings Institution.

Motor Vehicles Manufacturers Association (1987). *Motor Vehicle Facts and Figures '87.* Detroit, MI: The association.

Muller, T. ed. (1983). *Automated Guided Vehicles.* Berlin, Germany: Springer-Verlag.

Nakicenovic, N. (1986). "The Automobile Road to Technological Change: Diffusion of the Automobile as a Process of Technological Substitution", *Technological Forecasting and Social Change* 29: 309-330.

National Automobile Chamber of Commerce (1921). *Facts and Figures of the Automobile Industry.* New York: The chamber.

National Research Council (1984). *America's Highways: Accelerating the Search for Innovation.* Washington, D.C: Transportation Research Board.

National Research Council (1982). *Outlook for Science and Technology: The Next Five Years.* San Francisco: H. W. Freeman and Company.

Ogilby, R. B. (1924). *How Much Should I Spend For a Car?* Hartford, CN: State of Connecticut, Department of Motor Vehicles, Bulletin No. 45.

Pacey, A. (1983). *The Culture of Technology.* Oxford, England: Basil Blackwell.

Public Works Historical Society (1986). *One Hundred Years of Public Works Equipment.* Chicago: The society.

Sahal, D. (1980). "Models of Technology Development and Their Relevance to Advances in Transportation", *Technological Forecasting and Social Change* 16: 209-227.

Sahal, D. (1981). *Patterns of Technological Innovation.* Reading, MA: Addison-Wesley.

Santini, D. J. (1985). *Microeconomic Problems Arising from the Reshaping of Transportation Technology.* Argonne National Laboratory, Argonne, Il: Center for Transportation Research.

Strindt, F. A. (1983). *Matson's Century of Ships.* San Francisco: Privately published.

Stroble, H. ed., (1982). *Computer Controlled Urban Transportation.* New York, NY: Wiley.

U. S. Congress (1980). *Productivity: The Foundation of Growth*. Washington, D.C.: Joint Economic Committee.

U. S. Department of Commerce (1977). *Historical Statistics of the United States, 1776-1976.* Washington, D.C.: Bureau of the Census.

U. S. Department of Energy (1987). *Transportation Energy Data Book, No. 9.* Oak Ridge, TN: Oak Ridge National Laboratory.

U. S. Department of Transportation (1977). *America's Highways; A History of the Federal-aid Program: 1776-1976.* Washington, D.C.: Federal Highway Administration.

U. S. Department of Transportation (1987). *Highway Statistics: Summary to 1985.* Washington, D.C.: Federal Highway Administration.

Warnecke, H. J. ed. (1983). *Proceedings of the 2nd International Conference on Automated Vehicle Systems*. Amsterdam, Netherlands: North Holland Publishing Company.

Zimmermann, E. W. (1923). *Ocean Shipping.* New York: Prentice-Hall.

PART III

THE FUTURE OF SURFACE TRANSPORT

(a) Progress with Navigation Technology

Chapter 8

AUTOMOBILE NAVIGATION TECHNOLOGY: WHERE IS IT GOING?

Robert L. French
R.L. French & Associates, U.S.A.

8.1 INTRODUCTION

During the 1980s, numerous automobile and electronic manufacturers have become active in the development of sophisticated navigation, route guidance, and information systems that will provide unprecedented convenience and benefits to drivers in the 1990s and beyond. The potential of these systems for improving traffic management and for efficient and safe utilization of road transport has also prompted the formation of national programs such as the Caltrans/FHWA/General Motors Pathfinder Project in the United States, RACS and AMTICS in Japan, and international programs such as PROMETHEUS (see the APPENDIX) and other EUREKA projects in Europe. Both commercial and public sector organizations are already beginning to use some of the same technologies to monitor vehicle location and control fleets used for distribution and delivery of goods, courier and postal services, law enforcement, ambulance service, etc.

The term "road transport informatics" (RTI) has been suggested to describe the broad collection of functions and technologies involved (Svidén, 1987). In addition to navigation and route guidance, RTI includes automatic vehicle identification schemes for periodic billing of toll and parking fees, and systems for safe distance keeping, collision avoidance, and automatic chauffeuring.

Although automatic navigation and route guidance will be an obvious convenience for the individual driver, several studies have pointed toward important potential benefits for the motoring public as a whole. A recent economic assessment of the potential for improved motorist route following in the United States found that recoverable navigation waste amounts to 6.4 percent of all distance traveled by non-commercial vehicles and 12.0 percent of all time spent in such travel (King and Mast, 1987). The annual cost to individuals and to society of this excess travel was estimated at $45.7 billion considering only vehicle operating ($10.0 billion) and accident ($4.4 billion) costs and the value of time ($31.3 billion). Studies of the potential benefits of navigation and route guidance performed by the U.K.'s Transportation Road Research Laboratory point to similar potential benefits and advantages (Russam and Jeffery, 1986), as do the results of experiments and simulations in Japan (Fujii, 1986).

When a significant fraction of vehicles is outfitted with effective navigation aids, their diminished demands on roadway capacity will contribute to improved traffic conditions as well as economic savings. Automobile navigation systems with automatic routing or guidance functions are even more effective if current information on traffic conditions is available for consideration in routing. It has been estimated that the potential savings by eliminating navigation wastes in the United States increases from $45.7 to $73.6 billion annually if real-time traffic information is available to on-board route guidance systems (King, 1986).

While automobile navigation and route guidance concepts and their potential benefits have only recently received extensive and widespread attention, many of the underlying ideas are by no means new. This paper identifies background developments dating from early this century to the space age, and traces milestones leading to three distinct types of navigation systems now vying for positions in the RTI markets of the future. Each of these types, autonomous navigation systems, radio navigation systems, and proximity beacon navigation systems are discussed separately to highlight their strengths and weaknesses.

For maximum effectiveness, on-board systems, regardless of which type of navigation technology is employed, require infrastructure support to provide real-time data on traffic and road conditions, available parking, etc. Alternative communications links and means for supplying the infrastructure support thus required are also described.

What systems approaches and combinations of navigation and communications technologies will prevail between now and the year 2010? This paper concludes with a description of the author's impressions of "where automobile navigation technology is going", and a discussion of the rationale upon which the scenarios are based.

8.2 BACKGROUND

Automatic route guidance for automobiles was the objective of several ingenius mechanical devices developed in the United States around 1910 (French, 1987). Automatic vehicle location monitoring (AVLM) from a central location dates back to London Transport's Bus Electronic Scanning Indicator (BESI) deployed in the late

1950s (UMTA, 1981). The potential of on-board route guidance systems for improved traffic management has long been recognized, and was the major thrust behind the U.S. Federal Highway Administration's short-lived ERGS (Electronic Route Guidance System) project of the late 1960s (Rosen, et al., 1970), as well as the Japanese CACS (Comprehensive Automobile Control System) and West German ALI (Autofahrer Leitungs Information) projects of the 1970s (Gordon, et al., 1981). The ERGS, CACS and ALI concepts required infrastructure support in the form of roadside proximity beacons networked with central traffic computers to provide dynamic route guidance as well as to update vehicle location. These early experimental systems established patterns for contemporary proximity beacon approaches still being pursued in Japan, West Germany, and the U.K.

Microelectronics and advances in navigation technology have led to another type of route guidance and information system which has autonomous navigation capabilities and requires a communications link with the infrastructure only if real-time traffic information is to be used in conjunction with route guidance. The auxilary communications link may thus be in the form of area broadcasts from sparsely located transmitters rather than short-range proximity beacons spaced along the roadway. Autonomous systems use dead reckoning to determine approximate vehicle location, and map matching to correlate vehicle path with the road network as defined in an on-board map data base. This approach was invented and first applied to route guidance in the early 1970s (French and Lang, 1973), and is the technological basis for numerous systems that have been developed during the 1980s.

Also vying for RTI roles are navigation systems which employ radiolocation techniques. Dating from the 1940s is Loran, a hyperbolic navigation system using multiple pairs of ground-based master-slave transmitters. The current version, Loran-C, is used by most AVLM systems currently on the market in the United States.

The Transit satellite navigation system deployed by the U.S. Navy in 1964 has seen limited use for AVLM and has been tested for automobile navigation. More recent satellite systems still in the development stage include the U.S. Department of Defense's Navstar Global Positioning System (GPS) and the privately-operated Geostar Radio Determination Satellite Service (RDSS). Navstar GPS has potential applications for automotive navigation as well as AVLM but, like Loran-C, requires dead reckoning backup to maintain continuous location information because usable radio signals are frequently not available in mountainous or urban areas due to obstructions or interference. Geostar (called Locstar in Europe) RDSS is intended primarily for location monitoring rather than navigation.

8.3 AUTONOMOUS NAVIGATION SYSTEMS

Autonomous navigation systems are self-contained within the vehicle, and are capable of performing useful functions even in the total absence of external navigation signals or information from proximity beacons and land- or space-based radio transmitters. Three categories of autonomous systems have appeared during the 1980s.

8.3.1 Directional Guides

From 1981 through 1986, several relatively simple directional guides were introduced in Japan (Radley, 1982) and West Germany (Gosch, 1986) but, because their capabilities were limited compared to more sophisticated systems on the horizon, most of them have been withdrawn from the market. Nonetheless, this category of system is worthy of description because it introduces simple dead-reckoning concepts to which other features may be added to develop autonomous electronic map and route guidance systems. Dead reckoning is also used in conjunction with proximity beacons and as backup to radionavigation systems.

The principal components of a directional guide include heading and distance sensors, a microprocessor, a keypad or other input means, and a display unit. The driver keys in the coordinates of the origin and destination (in one system, these are read from a special map using a light pen), the microprocessor computes the vector connecting the two, and the display indicates the "crow flight" distance and direction to the destination. As the vehicle travels from the origin, signals from the distance sensor (usually derived from the speedometer cable) and the heading sensor (typically a solid-state flux-gate magnetic compass) are input for dead-reckoning computations of the updated position of the vehicle and the corresponding distance and direction to the destination.

Dead reckoning is the process of determining a vehicle's location and heading relative to an initial position by integrating measured increments and directions of travel. A vehicle's current position in rectangular (x,y) coordinates is given by

$$x = x_0 + \int \sin \phi(L)\, dL$$

$$y = y_0 + \int \cos \phi(L)\, dL$$

where x_0, y_0 are the initial coordinates, dL is the distance increment and $\phi(L)$ is the vehicle heading associated with the distance increment.

The dead-reckoning sensors most commonly used in automobile navigation systems were developed long before the automobile itself (French, 1986). These include the odometer (~50 B.C.), the differential odometer (~300 A.D.), and the magnetic compass (~1200 A.D.). The differential odometer is essentially a pair of odometers, one for a wheel on each side of the vehicle. Real-time computer analysis of differential odometer signals yields information on heading changes as well as distance travelled.

Dead reckoning accuracy decreases continuously with distance travelled. For example, the vehicular odograph, a WWII military navigation system based on the odometer and magnetic compass, accumulated one mile of error per 50 to 150 miles driven (McNish and Tuckerman, 1947). Although present generation dead-reckoning systems are less expensive and far more compact, little gain has been made in accuracy. Thus dead-reckoning navigation systems require periodic reinitialization to a known position.

8.3.2 Electronic Maps

Dead-reckoning errors notwithstanding, a simple directional guide keeps the driver informed of distance and direction to the destination. However, since the vehicle is usually unable to travel the "crow flight" path, the usefulness of this information is diminished unless the driver has sufficient knowledge of the road network to devise a rational route to the destination. In some situations, this may mean driving away from rather than towards the destination at times. More advanced systems thus include a road map display that indicates the position of both the vehicle and the destination on the road network.

The first such system to be marketed was the Honda Electro Gyrocator in 1981 (Tagami et al., 1983). This system used a gas-rate gyro and odometer to compute and display the vehicle path on a CRT screen. A transparent map overlay of appropriate scale was used to show the road network. This pioneering system was soon superseded by systems which use road maps stored in the form of digital data.

With state-of-the-art systems, digital road maps may be automatically selected according to the vehicle's location, shown on a CRT screen or flat display panel, and zoomed to the best scale for a particular situation. Vehicle location as well as destination may be indicated by symbols, and roads and streets may be shown with various intensities or colors to indicate classification. The most recent dead-reckoning system to be introduced with a sophisticated electronic map display is the Electro Multivision available in Japan on the Toyota Crown (Shoji et al., 1988). The Electro Multivision, which includes color TV, radio with cassette player, and CD player, is the first system to reach the market with CD-ROM for storing nationwide road maps, highway guides and dealer information (Ishikawa et al., 1988). A single CD-ROM can store 550 MBytes of data, enough for digital maps for the entire country.

Although not used in the Electro Multivision, the on-board availability of vector-encoded digital maps enables map matching, a software pattern-recognition process which continuously correlates measured vehicle path with the road network and thus prevents the accumulation of dead reckoning error. In map matching, the pattern of the vehicle's path is analyzed as a sequence of vectors deduced from any of a variety of dead reckoning processes. As the vehicle travels, its measured vector sequence is continuously compared with the digital map. Each time a turn is executed whose sense, magnitude, and location closely approximates those of a mapped turn, the vehicle is presumed to be at the mapped location. The matching process thus removes any dead reckoning error accumulated since the last turn (French, 1986).

The first commercially available electronic-map navigation system based on dead reckoning augmented by map matching is the Etak Navigator™ marketed in California since 1985. The Etak system uses a flux-gate magnetic compass as well as differential odometer for dead reckoning, and uses 3.5-MByte tape cassettes to store digital map data approximately equivalent to two paper street maps (Honey and Zavoli, 1985). The vehicle's location relative to its surroundings is continuously displayed on a monochrome CRT map presentation which may be zoomed to different scales. A fixed symbol below the center of the CRT represents the vehicle position, and points to the top of the display indicating vehicle heading. As the vehicle is driven, the map scrolls and rotates about the vehicle symbol to maintain an orientation corresponding with the driver's view of the streets. The destination, which is input by street number and name or by street name, and nearby cross street, is shown on the Etak screen as a flashing star.

8.3.3 Route Guidance Systems

Accurate electronic map displays, while far more useful than simple directional guides, have two shortcomings as autonomous navigation systems. First, the driver must view a busy display which some critics consider to be a potentially dangerous distraction from the driving task. Second, it is up to the driver to devise the most appropriate route and then follow it to the destination. Both problems are solved by the addition of algorithms for devising the best route to an input destination and for issuing real-time route guidance instructions.

Real-time route guidance directs the driver turn-by-turn to the destination. Once the driver specifies destination and routing criteria (fastest, shortest, scenic, etc.), the route guidance software makes all navigation decisions, freeing the driver to concentrate on driving safely. Explicit route guidance may be in the form of spoken instructions, displayed symbols (e.g., arrows shaped according to the manoeuvre), and/or displayed text messages.

Real-time route guidance over preplanned routes by an autonomous dead-reckoning/map-matching system was first demonstrated in the 1970s (French and Lang, 1973, and French, 1974). The first autonomous system to demonstrate on-board generation of best routes as well as real-time route guidance was Micropilot, a 1981 prototype that used an Apple computer to calculate the quickest route between an input starting point and destination (Wooten and Brett, 1986). Digitized voice was used to give turning instructions as each junction along the route was approached based on position estimated from odometry alone.

Although none have yet reached the market, there are several recent examples of advanced autonomous route guidance systems. Philips' CARIN (Thoone and Bruekers, 1984) was the first developmental system to use CD-ROM for storage of digital map data. The vehicle navigation subsystem employs dead reckoning (differential odometer and electronic compass based on the magneto-resistance effect) augmented by map matching. The system includes a route-search algorithm and provides step-by-step route guidance. In the original test and demonstration version, a color CRT map display shows vehicle location relative to the surroundings, and synthesized voice instructions prompt the driver when operating in the route guidance mode. The latest version has a flat-panel display for symbolic route instructions in addition to synthesized voice, and includes a CRT map display as an option. Philips proposes to equip CARIN with an RDS (radio data system) receiver so that broadcast traffic information may be taken into account in route selection (Fernhout, 1987).

Blaupunkt's EVA (Pilsak, 1986) is an autonomous map-matching system which uses a differential odometer and includes route-search software to generate an optimum route to coded input destinations. Turns at intersections, lane changes, etc. are specified on an LCD in the form of simplified diagrams which show lane boundaries and arrows to indicate the path to be taken. Synthesized voice capability is included, and is used to confirm destination entries as well as to articulate turn-by-turn real-time route guidance instructions. The first version, which has been tested and demonstrated since 1983, stores a digital map of the test city in EPROM. An enhanced version under development uses CD-ROM for map storage. Production versions of EVA may include an RDS receiver if RDS becomes a standard for broadcasting traffic information.

Nissan's NAV (Tsunoda et al., 1986) also uses dead reckoning with map-matching autonomous navigation. Map data is stored in a 1-MBit ROM. Optimal routes are computed in two stages using a coarse network for the initial calculations, and then interpolating using a more detailed network. The route is traced on a CRT map display, and special screens guide the driver through intersections along the way.

8.4 RADIONAVIGATION SYSTEMS

Radionavigation, which is commonly used for aerospace and maritime navigation, is becoming of more interest for land vehicles with the advent of satellite positioning systems and with the expanding coverage of land masses by Loran-C. Radionavigation has the inherent advantage of providing absolute location information. Its major drawbacks are that radio signals may be blocked or reflected by man-made structures in urban areas and by rugged terrain, and that radio receivers are subject to interference from various sources. Another drawback is the present high cost of satellite receivers, but this problem should be ameliorated when the systems come into large-scale use.

8.4.1 Satellite

The Navstar Global Positioning System (GPS) has received most attention as a basis for automobile navigation. Although the present focus is on Navstar, the older Transit satellite system has also been considered. In addition, other satellite systems, such as Geostar, MOBILSAT, and INMARSAT offer, or may offer, vehicular location monitoring services as well as mobile data communications.

Transit: The Transit satellite navigation system, implemented by the U.S. Navy in 1964 and scheduled to be phased out in 1996 (DOD and DOT, 1986), has several satellites logitudinally-spaced in polar orbits at a height of approximately 1,075 kilometers to give worldwide, albeit intermittent, coverage. Each satellite transmits information which, in combination with Doppler analysis, permits calculation of receiver location by interative solution of a set of equations. Since a Transit satellite is not always in range, vehicle systems based on Transit must include dead reckoning for continuous determination of position between satellite passes.

Ford Motor Company demonstrated the integration of Transit satellite navigation with dead reckoning for automobiles with an operational system installed in the Concept 100 car in 1983. This system periodically obtained accurate position fixes from the Transit navigation satellites to update the primary location from the dead-reckoning subsystem. The dead-reckoning calculation was based on odometer signals and a flux-gate magnetic compass which was software compensated for magnetic field pertubation by the vehicle (Jarvis and Berry, 1984). The vehicle's position was tracked on a map displayed on a color CRT with touch-screen controls.

Navstar GPS: The U.S. Department of Defense's Navstar GPS, which is still in the implementation stage with 21 (18 plus three operating spares) satellites to be launched into 12-hour orbits by mid-1991 (DOD and DOT, 1986), has received considerable attention as a possible basis for automobile navigation. When the GPS satellite constellation is complete, any point on earth will always be within range of at least four Navstar satellites.

A GPS receiver could accurately determine its three position coordinates by analyzing the travel time of signals from only three satellites if the receiver's clock is precisely synchronized with the atomic clocks that time the satellite signals. However, given the timed signals from four satellites, the GPS receiver solves a system of four equations for its three position coordinates (P_x, P_y, P_z) and the bias (C_B) of its less precise quartz clock:

$$(x_1-P_x)^2+(y_1-P_y)^2+(z_1-P_z)^2=(R_1-C_B)^2$$
$$(x_2-P_x)^2+(y_2-P_y)^2+(z_2-P_z)^2=(R_2-C_B)^2$$
$$(x_3-P_x)^2+(y_3-P_y)^2+(z_3-P_z)^2=(R_3-C_B)^2$$
$$(x_4-P_x)^2+(y_4-P_y)^2+(z_4-P_z)^2=(R_4-C_B)^2$$

where x_n, y_n, z_n, are position coordinates of the nth satellite based on ephemeris carried by its signal, and R_n is the range to the nth satellite.

Locations may thus be determined to within approximately 10 meters using P-Code signals intended for government authorized applications. Less precise C/A-Code signals for general use will permit location determination to within approximately 100 meters. Differential GPS, which uses a broadcast correction derived from a local receiver at a precisely known location, can increase the C/A-Code accuracy to within a few meters.

Since 1983 General Motors Corporation has tested and demonstrated GPS and other navigation systems installed in an automobile (Dork, 1983). In addition to Navstar GPS, the systems include Loran-C, and a flux-gate compass and odometer sensor for dead reckoning. This test configuration permits the relative evaluation of three independent or integrated approaches. The few satellites currently deployed limit adequate GPS signal availability to approximately 4 hours per day. Loran-C is used at other times to demonstrate radiolocation along with the integrated map display. Both GPS and Loran-C automatically shift to dead reckoning when their signals are obscured or when an unreasonable vehicle location is indicated by GPS or Loran-C. While there is no projected date for any of these systems to be introduced as a production option, General Motors Corporation is continuing to evaluate and develop the systems into practical and affordable navigational aids.

Navstar GPS was also the basis for CLASS (Chrysler Laser Atlas and Satellite System), a navigation system included in a concept car displayed at the 1984 World's Fair in New Orleans (Lemonick, 1984). CLASS included a nationwide set of AAA maps stored in image form on a video disc, and software for automatically selecting and displaying on a color CRT the map area incorporating the vehicle's current location as indicated by a cursor.

Although GPS has good potential accuracy and will provide continuous coverage, auxiliary dead reckoning is required in automobile applications to compensate for signal loss due to shadowing by buildings, bridges, foliage, mountainous terrain, etc. Dead reckoning may be used as a secondary system to maintain location information in the absence of GPS satellite signals or, alternatively, dead reckoning may be used as the primary vehicle location method with GPS employed as the secondary system to make position updates when usable signals are available. For example, a Nissan system relies upon dead reckoning whenever satellite position data disagree by 2 km or more, or if two consecutive satellite measurements taken at one-second intervals exceeds the amount of vehicle movement indicated by dead reckoning (Itoh et al., 1986).

A Department of Transportation evaluation of GPS for land vehicles notes that, because of differing ellipsoidal reference systems, the task of relating GPS location to local maps is formidable (Mooney, 1986). Hence map-matching may be useful with GPS as well as with dead reckoning. In fact, one Japanese automobile navigation system, the MAPIX-III (Ikeda et al., 1987) already integrates dead reckoning, GPS and map-matching. GPS signals, when available, provide absolute location which is reconciled with dead reckoning and matched to specific roadways included in the map data base.

GPS has also been used in a trial program for generating road map data in a rural area of the state of Tennessee (Knoernschild, 1988). Position data were recorded while driving a vehicle equipped with a GPS receiver over approximately 1000 kilometers of road. Test results of this type are generally comparable in accuracy with conventional 7.5 minute topographical maps, and show a high incidence of new highway construction, road re-alignment and map errors, even on recently revised maps.

In late 1987, the Department of Defense announced a decision to seek Congressional approval of funds to expand the number of GPS satellites from 21 to 24 to eliminate periodic areas of geometrically degraded accuracy and to improve system reliability. Thus full implementation may be delayed from 1991 to 1994.

Geostar RDSS: Geostar (called Locstar in Europe) will be the world's first Radiodetermination Satellite Service (RDSS) network with nationwide automatic vehicle location monitoring and two-way digital communications from the same set of user equipment (Richards and Snively, 1986). Major system elements include a control and central processing center (ground station), two or more geosynchronous relay satellites, and user transceivers. Except for very large vehicle fleets which might justify operating their own ground stations, Geostar service requires additional communication links between the central control and processing center and the dispatch offices of individual user fleets.

When complete, two Geostar satellites will be used in a triangulation process to determine vehicle location. The process starts at the base station with an interrogation signal containing a digital signature code that addresses the message to one or a group of vehicles, plus a digital notation indicating the time the signal was sent. One satellite relays the signal to the vehicle's transceiver where it is retransmitted to both satellites which, in turn, relay it back to the ground station. Recognizing its original time notation, the ground-station computer calculates the time it took the vehicle's retransmission to reach the separate satellites. Since the satellite positions are known, the computer can determine the vehicle's distance from each and pinpoint its location. The computer then uses digital map data to translate position coordinates into real-world location for transmission to the fleet terminal.

Full-scale Geostar service is scheduled for 1989. In the meantime, limited location service is provided by using a single Geostar satellite to relay Loran-C determined vehicle locations to the ground station.

8.4.2 Loran-C

Although Loran-C has been used for decades for marine and aircraft navigation, and had often been considered as a possible basis for automatic vehicle location monitoring systems (UMTA, 1981), dependable cost-effective Loran-C receivers designed specifically for the hostile electro-magnetic environment encountered by land vehicles have become available only in the last few years. Examples of state-of-the-art Loran-C AVLM systems are reported by Janc (1984), Bronson, et al. (1985), and by Carter and Warburton (1985).

Loran-C AVLM systems, unlike the obsolescent electronic signpost approach to AVLM, have the advantage of not requiring infrastructure support other than the existing Loran-C transmitter chains operated by the U.S. Coast Guard which is now installing mid-continent chains to complete nationwide coverage in the United States. European coverage is presently limited to the Mediterranean area and northern part of Europe, including the recently introduced French-operated chain. A European Working Group, formed to continue Loran-C service after 1994 when the U.S. Coast Guard plans to discontinue overseas Loran-C operation following completion of the Navstar GPS system (DOD and DOT, 1986), is considering expanding European converage.

Each Loran-C chain of 3 to 5 stations transmits time-synchronized groups of pulses at approximately 100 KHz. The time difference between arrival of pulses from the master and each secondary station describes a hyperbolic line of position. A Loran-C receiver measures the time difference of two or more master-secondary pairs, and the intersection of the lines of position defines the receiver's location. In most cases time differences are transmitted from the vehicle to the monitoring dispatch office for conversation to location coordinates which are used to show vehicle location on a CRT map display. Most systems include driver-operated switches for reporting status as well as location.

Land mobile radios, which many fleet operators already have installed for voice communications, are often fitted with modems to transmit vehicle location data to the dispatch office. However, at least one Loran-C based AVLM system uses cellular telephone (Carter and Warburton, 1985).

With the availability of high-performance Loran-C receivers at moderate cost, the types and number of AVLM installations have proliferated in the United States. In addition to traditional transit bus and police fleet monitoring, applications now include public utilities, waste management, security, and general trucking. Although Loran-C has not yet been seriously pursued as a basis for automobile navigation, the experience base being acquired through AVLM, along with the expanding geographical coverage, make it probable that Loran-C automobile navigation systems will also be developed within the next few years. However, like Navstar GPS, dead-reckoning backup and possibly map-matching augmentation will be required.

8.5 PROXIMITY BEACON NAVIGATION SYSTEMS

The proximity-beacon approach to vehicular navigation uses strategically located short-range transmitters (also know as "electronic signposts"), and the very reception of their location-coded signals indicates the receiving vehicle's instantaneous location. In route guidance applications, proximity beacons transmit routing and traffic information as well as location, and may include two-way communications with equipped vehicles.

8.5.1 Vehicle Location Monitoring Systems

The widest application of proximity-beacon technology is for AVLM systems such as those used for monitoring the location and status of transit buses from a central dispatch office (UMTA, 1981). In this type of application, an on-board system receives and stores a location code as the vehicle passes a proximity beacon. Upon periodic polling, the last beacon location and possibly the distance or time since passing the beacon are automatically radioed to the dispatch computer.

8.5.2 Route Guidance Systems

Several variations of the proximity-beacon approach have been investigated for interactive route guidance (Gordon et al., 1981). Typically, the driver enters a destination code for automatic transmission to a roadside unit as an equipped vehicle approaches instrumented intersections. The roadside unit, which may be networked with a traffic management system, analyzes the destination code and instantly transmits route instructions for display on the vehicle's panel. Alternatively, the roadside unit may only transmit its location to the vehicle where an on-board computer, using stored road network data, will generate instructions for continuing the route from the identified location. The proximity-beacon approach, inactive in the U.S. since the demise of the ERGS project (Rosen et al., 1970), has undergone further development in West Germany, the U.K. and Japan.

ALI-SCOUT: A state-of-the-art example of proximity-beacon route guidance systems is ALI-SCOUT, a joint project of the Federal Replublic of Germany, Siemens, Volkswagen, Blaupunkt and others (von Tomkewitsch, 1986). ALI-SCOUT combines certain characteristics of autonomous systems in that, while dependent upon proximity beacons, the in-vehicle equipment includes dead reckoning and map-matching featues that permit autonomous navigation between beacons which, consequently, may be spaced at greater intervals.

The ALI-SCOUT vehicular equipment receives approximately 8 KBytes of area road network data and recommended route data when passing strategically-located infrared proximity beacons. Simplified graphic driving directions to the input destination are presented in real time on a dashboard LCD. As an equipped vehicle passes each beacon, it transmits to the beacon stored data on its travel history since passing the last beacon. The equipped automobiles thus serve as traffic sensors. ALI-SCOUT will undergo large-scale field testing in the LISB (Leit- und Information System Berlin) project in West Berlin starting in 1988. Beacons are being installed at 20 percent of the traffic lights, and 700 equipped automobiles will be used in the tests.

Autoguide: In 1986, the U.K. Department of Transport proposed Autoguide, a proximity-beacon project starting with an early demonstration of interactive route guidance and traffic management in London (Jeffery, et al., 1987). Autoguide is a system for helping drivers find their way through the primary road network. A route computer is mounted in the vehicle and the driver enters the destination. Either visually or using synthesized speech, the computer then gives easy-to-follow instructions during the journey.

The on-board computer communicates with infrared beacons near main junctions, and the beacons are in communication with a central traffic information center. As an equipped vehicle passes each beacon, it transmits its destination, type and, possibly, preference for the type of route - the driver might want the quickest or the shortest (the two are not necessarily the same), or have some other requirement such as "no motorways". The beacon immediately transmits back to the vehicle directions to be taken at the junction, and the route computer translates them into simple instructions for the driver. Drivers are thus given guidance based on up-to-date information about the entire urban area.

While it is to be implemented with microelectronics and other advanced technologies that were not available 20 years ago, the Autoguide concept is quite similar to that of the Federal Highway Administrations' ERGS project of the late 1960s (Rosen et al., 1970). In fact, the original Autoguide planning was based on similar inductive loops buried in the roadway for the road-to-vehicle communication link. However, the Autoguide proximity beacons are now planned to be of the infrared type like ALI-SCOUT. Infrared beacons have the advantage of being less expensive as well as having higher data communications rates.

RACS: In Japan, the vehicle navigation systems approach currently promoted by the Ministry of Construction (MC) through the Road/Automobile Communications System (RACS) program depends upon proximity beacons for communicating traffic data and other information to on-board systems (Takaba et al., 1988). The RACS program, which is being coordinated by the MC's Highway Industry Development Organization (HIDO) and carried out by PWRI (Public Works Research Institute) in conjunction with a number of industry participants, is essentially an outgrowth of the earlier Ministry of International Trade and Industry's (MITI) CACS project (Fujii, 1986). Although the vehicular systems of several manufacturers participating in the program have autonomous navigation capabilities based on dead reckoning and map matching, proximity beacons are used for updating vehicle location as well as for communications. More information about this program is described in Chapter 9.

In 1986, PWRI started a 3-year cooperative investigation with preliminary transmission experiments, using both inductive radio and microwave type beacons. The first navigation field experiments were performed in March 1987 using radio beacons (247.2 KHz, 9600 bps) in conjunction with a digital map of 350 square kilometers, including parts of the Tokyo and Yokohama districts, prepared by PWRI. The experiment involved different in-vehicle navigation systems developed by private companies including Sumitomo Electric, Nissan, NEC Corporation, Nippondenso, Matushita, Mitsubishi, Mazda, NEC Home Electronics, and Yazaki.

Preliminary results indicated that further consideration of beacon power (range) is necessary, and that the digital map data base accuracy should be enhanced to include local roads, declination of terrestrial magnetism, and positions of selected facilities. Location errors between beacons resulted primarily from magnetic direction sensor inaccuracy, and map-matching was found to be very useful in reducing location error. PWRI is revising the digital map for the further field experiments scheduled for 1988 which will include proximity beacon transmission of dynamic information on traffic congestion.

8.6 INFRASTRUCTURE REQUIREMENTS

Infrastructure support requirements for automotive navigation and route guidance systems range from none for autonomous systems that provide only navigation or static route guidance, to intensive support for interactive route guidance systems based on the proximity-beacon approach. The infrastructure for different types of automotive navigation and route guidance systems may be differentiated according to that needed for the location or navigation process and that needed for dynamic (i.e., traffic responsive) route guidance.

8.6.1 Autonomous Systems

Autonomous systems, by definition, require no infrastructure support for the basic location and navigation functions, including route guidance based upon static or historical information. However, for maximum route guidance effectiveness, it is highly desirable for autonomous navigation systems to have access to information on current traffic conditions in the area being travelled. This, in turn, requires infrastructure support of two types. The first is for collecting information on traffic conditions and the second is for communicating this information to the on-board systems.

In many urban areas, extensive traffic sensing and reporting means are already in place for collecting real-time traffic data at central control centers for use in timing traffic signal lights, driving variable message signs, etc. Such systems may be expanded, and software added for analyzing traffic data to extract information for frequent transmission to equipped vehicles. More advanced systems would also maintain a data base of parking space availability, and information on the hours of service stations and other facilities of particular interest to motorists.

One of the most popular means under consideration for transmitting such data to autonomous navigation systems aboard vehicles in Europe is the Radio Data System (RDS) which uses side-band capacity of existing FM radio stations. RDS has the advantage of low implementation costs and of not requiring additional frequency allocations. Car radio receivers may be readily adapted to receive and decode RDS transmissions for input to the navigation system. General RDS test broadcasts are underway in several European countries, and RDS experiments in the Netherlands already include transmission of traffic data (Fernhout, 1987). The PROMETHEUS program is also investigating the possibility of satellite systems for monitoring traffic and broadcasting traffic data (PROMETHEUS, 1987).

Another communications approach that is suitable for autonomous navigation systems is the teleterminal being promoted in Japan by the AMTICS project (Okamoto, 1988). AMTICS (Advanced Mobile Traffic Information & Communications System), which has the objective of providing on-board traffic and navigation information by integrating a variety of subsystems, was established by the National Police Agency (NPA) through the Japan Traffic Management and Technology Association in cooperation with the Ministry of Posts and Telecommunications (MPT) at the beginning of 1987. An experimental pilot system will be started in Tokyo this year, and the first commercial operation may be as early as 1990.

A display screen in each equipped car will show traffic information gathered at control centers and broadcast to the cars using a mesh of cellular-like teleterminals each with a range of approximately 3 km. "Packet" data transmission technology will be used, and there will be capability for addressing messages to individual vehicles. Operation is at 800 MHz with data rates of 4800/9600 bps.

A demonstration of the use of in-vehicle traffic information in conjunction with autonomous navigation systems has been planned for the Pathfinder Project to be included in the Los Angeles "Smart Streets" Project. Pathfinder is a joint undertaking of the California Department of Transportation (Caltrans), the Federal Highway Administration, and General Motors. Traffic data already collected by conventional means will be augmented and communicated (probably via cellular telephone) to approximately 25 vehicles equipped with the Etak Navigator (or equivalent) described elsewhere in this paper.While the Etak system does not provide route guidance 'per se', software modifications will superimpose real-time traffic information on the electronic map display so that the driver can take it into account in route planning. Consideration is also being given to using the communication link to relay recent travel experience of test vehicles to the traffic data center for additional information on traffic conditions.

8.6.2 Radionavigation Systems

In most cases, the availability of radionavigation signals depends upon land- or space-based navigation transmitter facilities operated by government entities. In the United States, radionavigation policy and planning is established by the Department of Defense (DOD) and the Department of Transportation (DOT), and is promulgated by the Federal Radionavigation Plan (FRP) which is updated every two years (DOD and DOT, 1986).

Other than the obvious requirement for external radionavigation signals, the infrastructure support requirements for radionavigation systems are identical to those discussed above for collecting and transmitting real-time traffic data to autonomous navigation systems for dynamic route guidance. Thus it is reasonable to presume that the direct infrastructure costs for radionavigation systems are the same as those for autonomous systems provided no additional costs are incurred for the land- or space-based transmitters that, in general, are already required for air and marine navigation.

8.6.3 Proximity Beacon Systems

Infrastructure requirements are heaviest for proximity beacon navigation and route guidance systems because it is necessary for equipped vehicles to pass beacon locations at frequent intervals as they travel about. In addition to the usual real-time traffic data collection requirements outlined above for dynamic route guidance using autonomous and radionavigation systems, dynamic route guidance with proximity beacons requires wire-line or other communication links between the beacons and the central traffic data system. Thus, including the beacons themselves, the infrastructure expense is far greater for proximity beacon systems than for autonomous systems.

Nonetheless, conventional wisdom has traditionally held that route guidance systems which distribute most of their intelligence at the roadside rather than aboard individual vehicles leads to the minimum overall system cost because there are far more vehicles to be outfitted than there are junctions to be instrumented with beacons. By assigning more functions to the infrastructure side, the complexity and cost of on-board equipment is reduced. In particular, it is not essential that a large digital map data base and associated storage media reside aboard each vehicle if best routes are generated by a central traffic computer.

Conversely, if the on-board component of a navigation and route guidance system has a high level of capability with accompanying higher costs, the infrastructure side requires less capability and costs less. The minimum infrastructure cost would be that required to collect and communicate traffic data to on-board systems that are otherwise autonomous in operation. At the same time, this would result in the highest overall system cost because of the large numbers of vehicles to be equipped with the more expensive autonomous systems.

However, it is noted that designing for the lowest overall system cost is not necessarily the most effective way to proceed because this incurs heavy infrastructure expense which must be paid out of public funds unless some kind of user fee scheme is devised to transfer this expense to the owners of equipped vehicles. Several questions may be asked:

* Would vehicle owners prefer paying such fees on a continuing basis rather than buying more expensive equipment for their vehicles?
* Would the requirement for extensive installation of proximity beacons unduly retard the introduction and acceptance of route guidance?
* Would extensive infrastructure installations, once in place, hinder the future upgrading of equipment and services?

If vehicle units have no independent capability in the absence of proximity beacons, there will be no incentive for their purchase before sufficient infrastructure is in place, and even then the units will be useless when driving outside of instrumented areas. Upgrading of in-vehicle equipment would tend to occur gradually as vehicles are replaced, whereas upgrading of infrastructure equipment would probably be at infrequent intervals once the installations are in place. The argument for intelligent infrastructure rather than intelligent vehicles is further diminished by the continuing downward spiral in the cost of microelectronics aboard individual vehicles.

8.7 FUTURE SYSTEMS

The following scenarios describe the author's forecast of automobile navigation and route guidance developments for the next twenty or so years (i.e., until approximately 2010), and discuss the rationale upon which the projections are based.

Autonomous vehicular navigation systems will gradually come into widespread use between now and the year 2000. These systems will use digital maps for map matching to enhance dead-reckoning accuracy and for computing optimal routes for automatic route guidance. Some systems will optionally display digital maps for orientation and reference purposes. Lack of widespread availability of standardized digital maps will slow the market penetration of autonomous systems for the next few years.

The hardware/software architecture of the autonomous systems will accommodate the addition of appropriate receivers to take advantage of whatever type of traffic data communications scheme is available in the country or region in which equipped vehicles operate. Possibilities include RDS, teleterminals, and satellites, as well as adaptations of the cellular telephone. Although proximity beacons (radio frequency, microwave, infrared, etc.), will give way to map-matching for position updating, and to radionavigation for reinitialization, the proximity beacon concept is sufficiently well entrenched that it may persist in some areas as a first generation communications link. However, by 2010, standardized infrastructure to provide traffic data for dynamic route guidance will be in place in most high traffic density regions of the world, and the majority of new vehicles will be equipped to receive this information.

The autonomous systems will also be able to use optional radionavigation receivers to avoid the need for manual reinitialization (i.e., telling the system where it is) on infrequent occasions when dead-reckoning anomalies or erroneous map data confuse the software. Navstar GPS will be universally available, and Loran-C will be an option in regions where signals are available. As radionavigation receiver prices are reduced through volume production, they will begin to appear as standard equipment. But even then, the basic navigation systems may be characterized as autonomous because they can operate effectively in the event radionavigation signals are not available, as well as in areas where traffic information is not collected and broadcast.

The principal argument for the radionavigation role becoming secondary in automobile navigation and route guidance systems (i.e., for reinitialization) is that once all the features required to make a radionavigation system perform well under all circumstances are added, the satellite or Loran-C receiver could be removed and the system would continue to perform quite well until reinitialization is required. Specifically, dead reckoning is required because useful radionavigation signals are not always available, particularly in urban areas. On-board digital maps are also required to relate location coordinates to the real world because location coordinates, however accurate, are of little value to the driver. In addition, on-board map data are required for route generation and route guidance. With these elements in place, only software additions are required to match vehicle location with the digital map, thus overcoming minor map errors as well as location errors.

In contrast, radionavigation will continue to expand its present primary role in AVLM systems during the 1990s. The rationale for this projection is that neither continuous and precise map-matched positions nor on-board digital maps are required for most conventional AVLM applications. Loran-C systems will dominate these AVLM markets until sophisticated map-matching systems for navigation and route guidance become generally available as the year 2000 approaches. At that time, these more advanced systems will become the primary basis for AVLM in urban areas so that fleet management will have the added benefits of navigation and route guidance. However, satellite systems will continue to play an important role in WAVM (wide area vehicle monitoring) where they are particularly needed for fleet communications.

Although details of the above scenarios are based upon the author's personal views, the overall thrust and timing is generally consistent with the results of two recent Delphi studies. An International Institute for Applied Systems Analysis (IIASA) panel projected that autonomous vehicle navigation will be used by the majority of all commercial vehicles before the year 2000, and by the majority of all vehicles by 2010 (Svidén, 1987). The same panel of international experts projected that interactive route guidance will be introduced in 1995, used by the majority of commercial vehicles by 2005, and by the majority of all vehicles by 2010. A panel of U.S. Federal Highway Administration managers projected that assuming a healthy economy consistent with the IIASA assumptions, interactive route guidance will be introduced in 2001, and be used by the majority of commercial vehicles by 2011 (FHWA, 1988).

REFERENCES

Bronson, R., W. Sears and L. Cortland (1985). "II Morrow's LORAN-C Based Vehicle Tracking System", *Proceedings, RIN Conference on Land Navigation and Location for Mobile Applications:* Paper No. 14, York, England, September 9-11.

Carter, D. A., and R. D. H. Warburton (1985). "Using Cellular Telephones for Automatic Vehicle Tracking", *Proceedings, RIN Conference on Land Navigation and Location for Mobile Applications:* Paper No. 28, York, England, September 9-11.

Dork, R. (1983). "Automotive Navigation System", *Detroit Engineer & Michigan Engineering*, 44, No. 8, 11-13.

DOD and DOT (1986). "Federal Radionavigation Plan", *U.S. Department of Defense and Department of Transportation.*

Fernhout, H. C. (1987). The CARIN Car Information System and the Extension to CARMINAT", *Proceedings, Sixth International Conference on Automotive Electronics:* 139-143, London, England, October 12-15.

FHWA (1988). "Advancements in Vehicle and Traffic Control Technology", *Federal Highway Administration*: The Future National Highway Program 1991 and Beyond, Working Paper No. 7.

French, R. L., and G. M. Lang (1973). "Automatic Route Control System", *IEEE Transactions on Vehicular Technology*: VT-22, 36-41.

French, R. L. (1974). "On-Board Vehicle Route Instructions via Plasma Display Panel", *SID International Symposium Digest of Technical Papers:* 5, 146-147, San Diego, California, May 21-23.

French, R. L. (1986). "Historical Overview of Automobile Navigation Technology", *Proceedings, 36th IEEE Vehicular Technology Conference*: 350-358, Dallas, Texas, May 20-22.

French, R. L. (1987). "U.S. Automobile Navigation: Past, Present, and Future", *Proceedings, 16th International Symposium on Automobile Technology and Automation*: Vol. 2, 1-19, Florence, Italy, May 11-15.

Fujii, H. (1986). "The CACS Project and Now: Dynamic Route Guidance as the Final Target", *Annual Meeting of the Transportation Research Board*: Washington, DC, January 13-16.

Gordon, R. L., B. Soicher, and E. Donaldson (1981). "Study of the Feasibility and Design Configuration for In-Vehicle Route Guidance", *Federal Highway Administration:* Report No. FHWA/RD-056.

Gosch, J. (1986). "Smart Compass Pilots a Car to It's Destination", *Electronics:* 59, No. 21, 20-21.

Honey, S. K., and W. B, Zavoli, (1985). "A Novel Approach to Automotive Navigation and Map Display", *Proceedings, RIN Conference on Land Navigation and Location for Mobile Applications:* Paper No. 27, York, England, September 9-11.

Ikeda, M., H. Fukaya and T. Tabe (1987). "New Navigation System, MAPIX-III", *Proceedings, 16th International Symposium on Automotive Technology and Automation:* Vol. 2, 61-67, Florence, Italy, May 11-15.

Ishikawa, K., M. Ogawa, T. Tsujibayashi, and Y. Shoji (1988). "Digital Map on CD (Called CD Information)", *SAE Technical Paper Series*: No. 880221.

Itoh, T., Y. Okada, A. Endoh, and K. Suzuki (1986). "Navigation Systems Using GPS for Vehicles", *SAE Technical Paper Series:* No. 861360.

Janc, R. V. (1984). "Consideration of the Various Error Sources in a Practical Automatic Vehicle Location System", *34th IEEE Vehicular Technology Conference Record:* 277-284.

Jarvis, M. W., and R. C. Berry (1984). "Cathode-Ray Tube Information Center with Automotive Navigation", *SAE Technical Paper Series:* No. 840313.

Jeffery, D. J., K. Russam, and D. I. Robertson (1987). "Electronic Route Guidance by Autoguide", *TRAFFEX '87 Conference:* Brighton, England, April 7-9.

King, G. F. (1986). "Potential Benefits of Route Guidance", *Annual Meeting of Transportation Research Board*: Washington, DC, January 13-16.

King, G. F., and T. M. Mast (1987). "Excess Travel: Causes, Extent, and Consequences", *Transportation Research Record 1111:* 126-134, TRB, National Research Council.

Lemonick, M. (1984). "Now: Driving by Satellite", *Science Digest*: Vol. 92, 34.

Knoernschild, G. F. (1988). "Global Positioning System for Vehicle Navigation and Position Reporting", *Proceedings, 18th International Symposium on Automotive Technology and Automation:* Vol. 2, Paper No. 88080, Florence, Italy, May 20-June 3.

McNish, A. G., and B. Tuckerman (1947). "The Vehicular Odograph", *Terrestial Magnetism and Atmospheric Electricity:* 52: 39-65.

Mooney, F. W. (1986). "Terrestial Evaluation of the GPS Standard Positioning Service", *Navigation, Journal of the Institute of Navigation:* 32, 351-369.

Okamoto, H. (1988). "Advanced Mobile Traffic Information and Communications System (AMTICS)", *Annual Meeting of Transportation Research Board*: Washington, DC, January 11-14.

Pilsak, O. (1986). "EVA - An Electronic Traffic Pilot for Motorists", *SAE Technical Paper Series*: No. 860346.

PROMETHEUS (1987). "Systems Engineering and Standardization", *Topics of Research:* PROMETHEUS Working Group 10.

Radley, K. (1982). "Getting There...", *Motor:* March 6.

Richards, R. T. and O. Snively (1986). "Geostar Positioning Analysis", *Proceedings, IEEE Position Location and Navigation Symposium:* 13-19, Las Vegas, Nevada, November 4-7.

Rosen, D. A., F. J. Mammano and R. Favout (1970). "An Electronic Route Guidance System for Highway Vehicles", *IEEE Transactions on Vehicular Technology*: VT-19, 143-152.

Russam, K., and D. J. Jeffery (1986). "Route Guidance and Driver Information Systems - An Overview", *Proceedings, 2nd IEE International Conference on Road Traffic Control:* 1-5, London, England, April 15-18.

Shoji, Y., T. Horibe, and N. Kondo (1988). "Toyota Electro Multivision", *SAE Technical Paper Serices:* No. 880220.

Svidén, O. (1987). "A Delphi Panel-Derived Scenario on Road Transport Informatics Evolution", *International Institute for Applied Systems Analysis:* WP-87-32.

Takaba, S., M. Katakura, H. Kawashima, M. Shibata and D. Fujita (1988). "Experimental Study on Road/Automobile Communications System in Japan", *18th International Symposium on Automotive Technology and Automation:* Paper No. 88078, Florence, Italy, May 30-June 3.

Tagami, K., T. Takahashi, and F. Takahashi (1983). "Electro Gyro-Cator, New Inertial Navigation System for Use in Automobiles", *SAE Technical Paper Series*: No. 830659.

Thoone, M. L. G., and R. M. A. M. Breukers (1984). "Application of the Compact Disc in Car Information and Navigation Systems", *SAE Technical Paper Series*: No 840156.

Tsunoda, M., T. Itoh, S. Ishikawa, and M. Iwabuchi (1986). "Navigation Systems for Passenger Cars", *SAE Technical Paper Series:* No. 861056.

UMTA (1981). "Automatic Vehicle Monitoring Program Digest", *Urban Mass Transportation Administration:* Report No. DOT-TSC-UMTA-816-11.

von Tomkewitsch, R. (1986). "ALI-SCOUT - A Universal Guidance and Information System for Road Traffic", *Proceedings, 2nd IEE International Conference on Road Traffic Control*: 22-25, London, England, April 15-18.

Wootton, H. J., and A. C. Brett (1986). "Route Information Systems - Signposts of the Future?" *Proceedings, 2nd IEE International Conference on Road Traffic Control*: 15-17, London, England, April 15-18.

Chapter 9

THE CACS PROJECT: HOW FAR AWAY ARE WE FROM THE DYNAMIC ROUTE GUIDANCE SYSTEM?

Haruki Fujii
Association of Electronic Technology for Automobile Traffic and Driving, Japan

9.1 INTRODUCTION

The CACS project was carried out over 6 years by the Agency of Industrial Science & Technology, MITI, beginning in 1973. In this project, the dynamic route guidance system (DRGS) with considerable scale was experimentally implemented on the road network of Tokyo metropolitan area as the first challenge. Immediately after the project was completed, the Association of Electronic Technology for Automobile Traffic and Driving (JSK: Abbr. of Japanese name) was established by MITI and major private companies in order to follow up on the project.

The activities of the JSK foundation are to revise and extend the technological results interweaving recent micro-electronic technologies, and to expedite utilization of the results to various users, especially to road and traffic authorities.

Turning to the recent movement of this field, vivid R&D activities concerning navigation systems for automobiles are cultivating the process for future enhanced road-automobile information systems among which DRGS would be positioned as a final target, accompanying the development and introduction of various kinds of communication media for automobiles (represented by cellular telephone and RDS etc). However, many of the technological and administrative problems concerning this development process have not changed so much, in spite of the rapid progress of microelectronics in the latest decade.

This paper introduces the points of technological investigation required for implementing the functions of DRGS and discusses about the developing process toward enhanced road-automobile information systems, standing upon the experiences of CACS project and JSK's activities, and also intends to give some suggestion to decision makers in various fields with respect to the future of information service systems such as DRGS.

9.2 THE CACS PROJECT AND ITS FOLLOW-UP ACTIVITIES

9.2.1 The CACS Project

In Japan, the rapid progress of motorization began in the late 1960's, causing a number of social problems such as traffic congestion, traffic accidents, air pollution and so forth. In order to mitigate these problems, the CACS (Comprehensive Automobile Traffic Control Technology) project was carried out aiming to develop versatile information systems combining the technologies of various fields such as electronics, communication and traffic control.

The CACS Project was composed of the following 5 subsystems:

1. Route guidance subsystem
2. Route display board subsystem
3. Driving information subsystem (warning system by in-vehicle display)
4. Traffic incident information subsystem (a kind of roadside radio)
5. Public vehicle priority subsystem (study on software only)

In this project, the stress was put upon the research and development of the route guidance system. The system design was carried out assuming the realization of this system was to come 20-25 years later. Therefore the policy to introduce enhanced functions was challengingly adopted, as far as circumstances permitted.

As the final stage of the project, a large-scale experimental system (pilot test system) was constructed in a southwestern area of Tokyo and operated from October 1977 for over a year to evaluate total system performance.

9.2.2 Post-CACS Activities

The follow-up activities have been carried out by the JSK foundation under the control of MITI, aiming at the revision and pervasion of CACS results. These include the activity to study future information systems around automobiles in cooperation with related public organizations. Although the results of these activities are not directly adopted to the policies or undertakings of other public organizations, the way of thinking and experience of the study have made large contributions to the activities of various fields through the people who have participated in the CACS project and JSK's activities.

The activities of JSK are as follows:

Technological cooperation activities

The pilot test system built in the CACS project was transferred to the National Police Agency and the Ministry of Construction to be utilized as a tool through which future traffic control technologies could be investigated. JSK is continuing to give positive aid and support to the undertakings of these agencies.

Experimental Research Activities

a) development of digital communication technology between road and automobile, usable for various systems. (1980-) (to be mentioned later)
b) Research on car-to-car digital communication technologies (1981-)
c) Development of various test systems for the purpose of testing the results of R&D and proposing their applications. (The "Tsukuba Expo" 85 system was a major example, and was installed in order to test the revised inductive transmission technology.)

General investigation and research (1985-)

Trend investigation of future information systems for automobiles' taking account of socio-economic changes.

9.3 DYNAMIC ROUTE GUIDANCE SYSTEM IN THE CACS PROJECT

9.3.1 Configuration and Functions of the System

System configuration and test area

The pilot test system of CACS was built in the southwestern area (Approx. 30 km^2) of Tokyo shown in Figure 9.1.

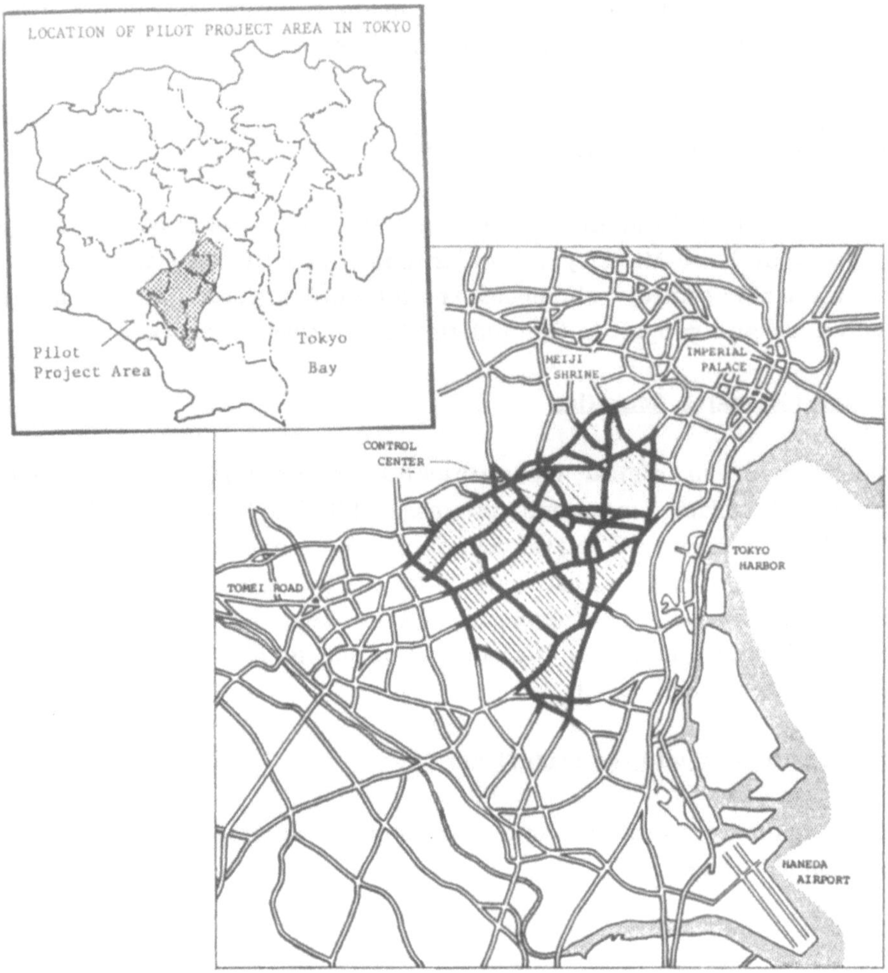

Figure 9.1 Pilot Test Area

Over 100 roadside units were placed at major intersections and connected with a central system (which was installed in the experiment center) through telephone lines. In-vehicle units and roadside units were linked by an inductive transmission system through loop antennae. Approximately 1300 in-vehicle units containing 1000 units used exclusively for data collection (without display function) were prepared. The whole structure of the system is shown in Figure 9.2.

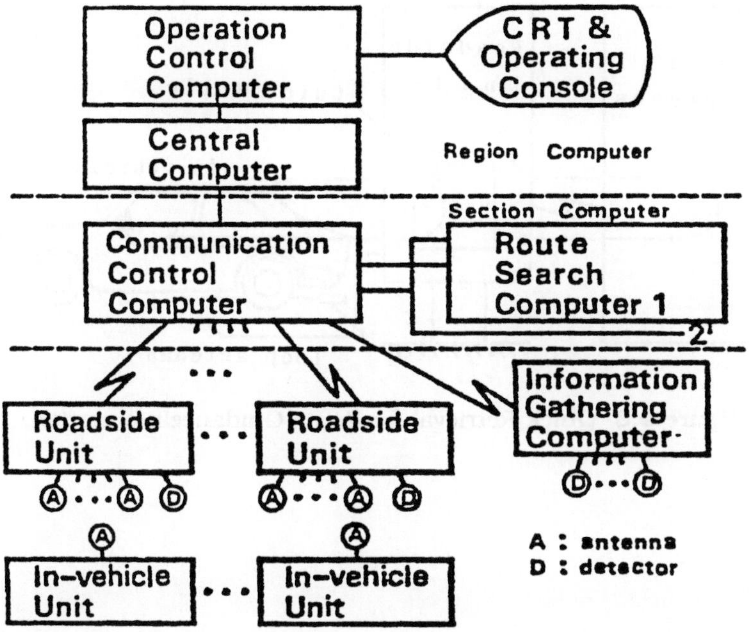

Figure 9.2 Configuration of the Route Guidance System (CACS).

Functions

- **Central system**

Using the records of communication which are gained when the cars equipped with in-vehicle units pass over loop antennae which belong to roadside units, the travelling times for every link of road network are estimated and forecasted. Then the shortest time routes for the combination of all points are calculated every 15 minutes. Furthermore, "guide tables" which are used to refer to guidance information (the directions to be selected at the intersections) for destinations being given, are computed nd sent to each roadside unit.

- **Roadside units and vehicle-units**

Before starting, 7-digit destination codes are set through encoders by drivers. During the time in which vehicles pass over loop antennae, roadside units receive destination codes, retrieve the guidance information concerned and return the results to the vehicle units (Figure 9.3). The guidance information is displayed as shown in Figure 9.4.

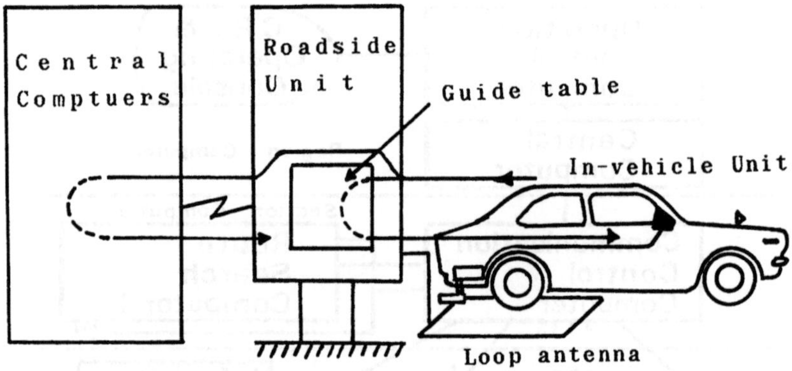

Figure 9.3 Quick Retrieval for Route GuidanceInformation.

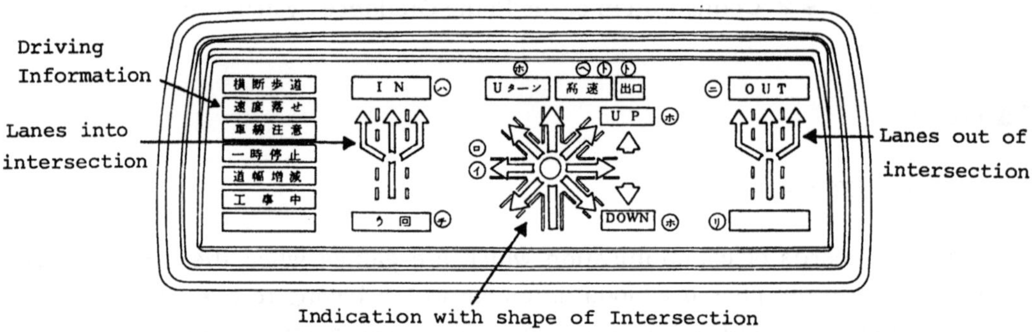

Figure 9.4 Display Pattern of In-Vehicle Unit.

9.3.2 Investigation About the Benefit of DRGS Through Simulation

The benefits and effect of DRGS upon traffic flow were studied through large scale simulation models and a pilot test.

Study by simulation model

Combining the traffic assignment model (1500 intersection network) and the dynamic traffic flow simulator (100 intersections, 20,000 vehicles), the following studies were carried out:

a) to estimate the benefit of traffic flow improvement - reduction of gross travelling time 3-9% (the economic benefit for Tokyo ward area was estimated at approx 20 billon yen/year)
b) to study the effect of DRGS upon traffic flow- the phenomena of traffic flow caused by shortest time route guidance (oscillation of traffic flow on alternative routes), the effect of traffic flow distribution, etc.

9.4 TECHNICAL ISSUES ASSOCIATED WITH THE IMPLEMENTATION OF DRGS

Major technical items which were studied in the design process of the CACS route guidance system are as follows.

9.4.1 DRGS and Communication Media

In order to realize the dynamic route guidance function, communication media through which information on the changing traffic situation is taken into in-vehicle units are required. In the case of CACS, a two-way inductive transmission (by weak power) system was adopted to fulfill the following conditions which were necessary to realize the route guidance functions as mentioned in the previous chapter:

a) to provide timely guidance informations for drivers (function to local current place accurately)
b) to respond to respective automobiles according to their destinations or requests (two way spot communication channel)
c) to be able to serve every automobile (restriction of resource of waves)

Points on how to fulfill these conditions are still to be discussed, even now when dead reckoning navigation systems with in-vehicle digital maps are about to be commercialized.

Relating to a), although the function to locate the current position can be implemented by technologies such as gyroscopes, utilization of satellites, map-matching technologies and their combination, the utilization of spot communication systems like "beacons" remains the major candidate.

Items b) and c) should be discussed in relation to each other. As a matter of common knowledge, the systems which provide navigation systems with information about congestion and traffic accidents through one way communication media such as highway radios, RDS (radio data system) and so forth, are now being studied in Europe and Japan. From the standpoint of traffic control, this would be a practical way to fulfill a large part of the expectation for DRGS.

On the other hand, it is necessary to take into account the situation when most drivers require more detailed information through inquiry systems. From this standpoint, the restriction of resource of electromagnetic waves becomes an important issue. As the communication media for this purpose, the cellular telephone system would be a major candidate. However attention should be given to the case when many inquiries caused by traffic congestion overwhelm the capacity of this system, because these kind of inquiries would simultaneously come from a limited area. Therefore the percentage of automobiles to receive enhanced information service would be a significant issue for discussion in respect of communication media between roads and automobiles.

As the communication demand for automobiles will possibly increase as a result of various information services required for convenient drives or communication services for fleet control, many kinds of communication media for automobiles are newly proposed and being developed in Japan. Among them, a small zone two-way digital transmission system using a multi-channel access system is supposed to be an important candidate. On the other hand, the inductive transmission system cannot be neglected due to a technical aspect of the simple transmission protocol.

As for the inductive transmission system, JSK has proposed that the specification be revised as follows:

1. to widen the choice of antenna types in order to adapt to the roads and roadside conditions in Japan;
2. to add location information into the data set which is transmitted from roadside antenna for navigation systems;
3. to provide the communication protocol with a device through which various systems can be served by only one channel;
4. to devote exclusive bytes in the data set from in-vehicle unit to traffic data acquisition for public use.

9.4.2 Generation of Traffic Information in the Route Guidance System

It is necessary to acquire information about the status of traffic flow in the road network concerned to carry out dynamic route guidance. The quantitative data are required to make it easy to compare the travel "cost" of alternative routes, setting aside the case of inevitable detouring due to traffic accidents and special events etc.

Although discussing the travel "cost" is a big problem itself, in the CACS project travel time was used. Travel time would be desirable information for the driver even when it is directly provided as traffic information. In the CACS implementation, the logs about the communication which took place between every roadside antenna and in-vehicle unit were accumulated and the times elapsed betwen two adjacent points were calculated (Figure 9.5).

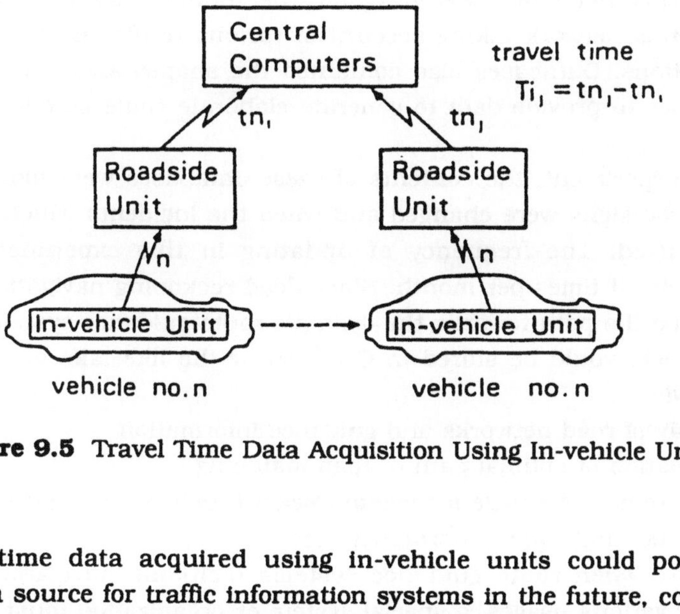

Figure 9.5 Travel Time Data Acquisition Using In-vehicle Units.

The travel time data acquired using in-vehicle units could possibly be an important data source for traffic information systems in the future, considering the progress of navigation systems. The route guidance system of CACS was so designed that the shortest time routes be updated every 15 minutes, taking account of the change in traffic patterns. If the route guidance information is calculated using only the original data, the shortest paths which are provided to drivers would be somewhat outdated, because it takes considerable time to put collected data in order, to calculate shortest routes and to transmit the results to roadside units. Furthermore it should be considered that some time will elapse before the automobiles reach the places where the results of the calculation are to be provided.

In the CACS project, the methods to forecast the future travel times using the varying patterns made from past records were studied, and part of the results was implemented in the route guidance system. Although travel time forecasting remains a difficult problem in spite of various efforts, effective methods might be developed in the future when we can utilize sufficient data concerning travel times, occurrences of trips, OD's and routes etc., because the increase of in-vehicle information systems and the development of inquiry systems will provide those data for the traffic information system.

9.4.3 A Database for Road Networks and their Management

The CACS route guidance system was implemented so that it worked upon a road network which was composed of about 70 major intersections, and databases were prepared for this network. Searching for the minimal travel time routes was performed on this network taking account of various traffic signs, restrictions and road constructions. Databases also contained the shapes and structures of intersections in order to provide data to generate elaborate route guidance instructions for drivers.

During the experiment, the contents of these databases were maintained at the time when traffic signs were changed and when the incidents which affected route selection occurred. The frequency of updating in this experimental area was restricted to several times per month. Since dead reckoning navigation systems are considered to be the path towards the dynamic route guidance system, the onboard data bases which would be stored in CD-ROM or the like take on very important roles; as follows:

a) for display of road networks and guidance information
b) for calculation of optimal path or map-matching
c) As the common language for communication between roadside and in-vehicle units concerning traffic information, etc.

In the future when route guidance systems including navigation systems are introduced for various usages, a special system or organization must be established to deal with the maintenance tasks of road network data bases. In the case of the CD-ROM type data base, as the period between revisions of them would be inevitably lengthy, it is necessary for road networks in CD-ROM to be used after some updating for recent changes.

Therefore the convenient circulating routes of these complimentary updatings should be prepared so that drivers can get them easily; for example at gas stations or through telephone or facsimile messages to their homes.

On the other hand, some devices might be required for the structure of data bases in order to make it easy to add complementary updating. Relating to this issue, the coding of "places" must be mentioned as an important problem.

9.4.4 Coding of "Place" and its Input Method

The Japanese way of addressing is different from that of America and Europe, who use the combination of street names and numbers. In Japan the addresses are expressed using the numbers of blocks surrounded by roads and partition numbers not chosen in a systematic way. It is therefore difficult to design the input method for destinations in Japan. Although it is possible to set a destination directly upon the map displayed on a CRT at present, in the CACS implementation we use dial switches for the input device.

In the CACS project, the "places" were defined to be the "links" between major intersections, and 7 digit destination codes were assigned to them. The network was expressed using this "link" concept (Figure 9.6).

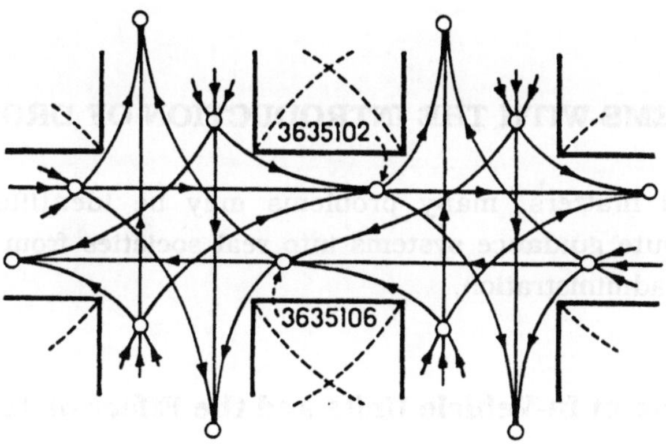

Figure 9.6 Expression of Destination and Road Network Structure.

This system was designed out taking account of other situations that we experienced when we were unable to reach our final destination crossing opposite lanes in Tokyo. Although this coding system was short of its capacity to express entirely precise road networks, it was inevitable for us to economise on memory due to the level of microelectronics at that time. In Japan, the input method of destinations tends to be troublesome, so this would possibly be the key for further progress of some route guidance systems or navigation systems.

9.4.5 Providing a Method of Route Guidance Information

In the CACS route guidance system, drivers were given not only the direction which they should select, but also the suitable lanes before or after crossing intersections by in-vehicle units. This display method was very useful when we had to drive in those cities which had intricate road networks like Tokyo, especially when geographical knowledge was inadequate. This success owes much to the exquisite timing of displays.

On the other hand, as we also know that this display method might be troublesome for the drivers who have much geographical knowledge, further investigation should be made on this aspect. However, it would possibly be difficult to realize sufficient guidance without CACS-type display methods in cities like Tokyo, which have many two-level crossings and complex intersections. Furthermore, room is left for providing methods for guidance information to be improved by applying audio input-output technologies.

9.5 PROBLEMS WITH THE INTRODUCTION OF DRGS

Beyond technical matters, many problems may be identified concerning the introduction of route guidance systems into real societies from the view points of management and administration.

9.5.1 Pervasion of In-Vehicle Units and the Effect of Traffic Information Service

By the simulations mentioned previously, we studied the phenomena presumed to occur in traffic flows by introduction of DRGS and counterplans to them. In the case where shortest time route guidance is adopted, when the ratio of the vehicles equipped with in-vehicle units exceeds a certain threshold, traffic flow begins to oscillate between alternative routes. The method which distributes traffic flow to the former shortest time routes in a certain ratio effectively copes with this phenomenon and by this method we are able to reduce gross travel time continuously.

However, when it is supposed that in-vehicle units are to be purchased voluntarily, the number of vehicles which are equipped with in-vehicle units increases. Although the people who paid for vehicle units can benefit when the equipped vehicles are small in number, as the number of equipped vehicles increases, the benefit to those who paid earlier would be lessened comparatively, because the increase in number would improve the traffic flow condition generally, and it would give some benefit to non-equipped vehicles. (As a matter of course, the benefit of geographical knowledge remains with the vehicles equipped with in-vehicle units.) Presuming that the pervasion of the in-vehicle units such as automobiles' navigation systems would be promoted by the private sector, we might reach a point where the pace of the pervasion begins to slow down, because the incentive for purchasing in-vehicle units would be based upon the advantage of faster arrivals.

9.5.2 The Future Relationship between Traffic Control Systems and Information Service Industries

It is presumed that various sectors or industries could take part in the development process toward the DRGS, and then it will become important to adjust their roles in the future. At present, in order to take the future method of traffic control into account, such systems serving digital encoded traffic information to navigation systems through broadcasting systems like RDS or radio communication channels, are being developed in various countries. These efforts seem to spring mainly from the public sector. If the navigation systems become the components of traffic information service systems, a higher service level of traffic information will be required; being pushed by the expectation of drivers and the tendency of private companies to fulfill this need.

Meanwhile the demand for better traffic information from business vehicles is higher than that of general drivers.

In the transport industries in Japan, we sometimes find that firms try to get traffic information at a cost or acquire information using their own communication networks. These movements suggest that the traffic information service which can reply to the respective demands of drivers will be introduced before long. In this type of system, there is supposed to be room for the communication and information industries to participate. So it will become an important problem to discuss the role of the private sector in relation to the public sector for traffic control.

9.5.3 Initial Investment in Infrastructure and the Possibility of Fleet Control Systems

When we tried to realize the CACS type DRGS system, the first question to be answered is how to prepare the initial infrastructure which has sufficient scale and service to start up the pervasion of in-vehicle units. Presuming that the infrastructure should be prepared by the public sector, the difficulty depends not only upon the largeness of the scale of investment, but also on the uncertainty of its necessity caused by the fact that the pervasion of in-vehicle units is not a confirmed process. However, in several major cities of Japan, bus-fleet control systems which are composed of CACS type transceiver (inductive transmission) infrastructure and in-vehicle units have already been introduced with a subsidy from the Ministry of Transportation. These systems are introduced to improve services to passengers, and reduce personal expense through automating fleet managing tasks. As other kinds of transport industries also require the fleet control functions, thus there is a possibility to extend them to general purpose communication infrastructure for various kinds of vehicles. If this possibility is brought to realization, we can expect further functional developments associated with navigation systems.

9.6 CONCLUSION

As is mentioned above, various problems must be discussed and resolved before enhanced information systems for automobiles such as DRGS are brought into operation in our society. As it is most likely that navigation systems and traffic information systems will make progress in cooperation with each other, most of these problems should be tackled by cooperation between the public and private sectors. The management of traffic information, should basically be put under the control of public sector, but some value-added services by the private sector ought to be permitted to some extent, if the development and pervasion of enhanced in-vehicle information systems are to arouse the efforts of the private sector.

Another major point concerning the information service system, which requires roadside units with a communication function is who invests in the roadside facilities and who operates them. It tends to be assumed that investment for this kind of infrastructure should be made by the public sector, because of that sector's close relationships with traffic control systems associated with this infrastructure. An unlikely scenariois for this investment to be made by the private sector such as the communication industry, transport industry and others.

We at JSK have begun a case study to investigate the feasibility of this way of thinking in the eastern part of Shizuoka prefecture.

REFERENCE

Kobayashi, F. and Izumi, M. (1977). "Comprehensive Automobile Traffic Control Project -- Part II Improvement of Traffic Flow by Route Guidance System", paper 9, 3rd Automotive Engineering conference (SAE of JAPAN and SAE of Australia).

Agency of Industrial Science and Technology, Ministry of International Trade and Industry, ed. (1979): R&D of the Comprehensive Automobile Traffic Control System (in Japanese).

Chapter 10

CONTRIBUTIONS OF TRANSPORTATION NETWORK MODELING TO THE DEVELOPMENT OF A REAL-TIME ROUTE GUIDANCE SYSTEM

David E. Boyce
University of Illinois at Chicago, USA

10.1 INTRODUCTION

The choices facing most urban travelers concerning frequently - made trips, such as the daily journeys to and from work, pertain to the route of travel and the time of departure. Some of these travelers may also consider the choice of mode, if indeed a choice is available to them. The basis for these choices largely consists of the traveler's experience with the performance of specific travel choice combinations selected during past excursions. Even this experience, however, is gained sequentially as a series of related events; real opportunities to compare the performance (e.g. travel time, cost, comfort, etc.) of alternative choices of routes and departure times are rarely available. Since travel conditions vary markedly from hour to hour, as well as day to day, this sequential experience provides only a rough guide to the traveler, and offers little opportunity for ex post comparison with discarded choice alternatives.

For many years, some general information on travel conditions has been available in most large urban areas in the form of traffic and weather advisories. Such information can be helpful in avoiding travel delays caused by specific incidents, such as accidents, flooding, or snow and icy conditions at a location affecting hundreds of travelers. On a normal weekday, however, these advisories serve only to confirm to travelers that they should rely on their own past experience.

Recent technological developments in vehicle navigation suggest that the future situation could be quite different. Travelers may be able to obtain in advance, as well as en route, estimates of the expected performance of the best route of travel, as well as alternative routes they may specify (e.g. avoiding a specific bottleneck). Moreover, they may be able to obtain these estimates for several departure times, such as every ten minutes for the next hour. Furthermore, this information may be available for public transit services as well as private automobiles, thereby providing a more informed basis for choice among these modes.

If such a capability were implemented, a new problem would arise which is solved implicitly by the present experienced-based system: how should travelers be allocated to the several competing routes from an origin area to a destination area? If all such travelers were advised to take the same route, the traffic congestion produced might well be much worse than without the vehicle guidance system.

This problem has not escaped the attention of past developers of vehicle guidance systems. For example, in 1970 the designers of the electronic route guidance system (ERGS) of the U.S. Department of Transportation cautioned:

> "Efficient algorithms, which can handle large networks, and the great amounts of data thus implied, will have to be developed. Strategies must be developed to route only the correct proportion of drivers via parallel routes... Also envisioned is a methodology for acquiring origin and destination data from roadside units and forecasting real-time demand on network links. Demand would be compared to capacity, and routing strategies would operate to maintain traffic flow below the link breakdown volume... Obviously, much research is still required to develop the software that would operate such a sophisticated information and control system" (Rosen et al., 1970, p. 152).

In making this statement, the authors suggest that their hopes for algorithmic and computer advances may have even exceeded their hardware design ambitions, which anticipated the current generation of micro-electronics and related advanced technology required for its implementation. In this regard, it is interesting to note that the first algorithm of the type described was only invented in 1968 (Dafermos and Sparrow, 1969), based on a problem formulated in the 1950s (Beckmann et al., 1956), and was not applied in planning large networks until the late 1970s (Eash et al., 1979).

Likewise, the developers of the Comprehensive Automobile Traffic Control System (CACS) project carried out in Japan since 1973 questioned the effectiveness of a dynamic route guidance system:

> "If we use a simple scheme such as shortest-time path guidance (when most of the vehicles are under control of the system), it might cause oscillation of traffic flow among alternative routes. In order to avoid this situation, an advanced scheme such as route distributing guidance should be prepared" (Fujii, 1986, p. 2).

This phenomena was studied by CACS using computer simulation; oscillation of traffic flow was observed in the simulations when the proportion of vehicles under route guidance control exceeded 50%. Nevertheless, in a pilot test of dynamic route guidance implemented in Tokyo in 1977, travel time was reduced by 11% on the average by using shortest-time routes. In this test, the proportion of guided vehicles was very small.

From a transportation network modeling perspective, it is interesting to observe that the developers of these systems seemed unaware both of the complexity of the routing problem and the potential contributions of network modeling to solving it. While their statements concerning the need for multiple route guidance reveal an intuitive appreciation that multiple routes are used in congested transportation networks, they also display an apparent naivety with regard to the difficulty of devising a multiple route guidance system. As I have become more familiar with current developments in vehicle navigation (French, 1987; Dial, 1987; Boyce, 1988), I realize that little has changed with regard to this observation. Therefore, I felt a chapter on this perspective offered a meaningful contribution to this book.

The chapter is organized in the following way. First, the current vehicle navigation technology is reviewed from a network routing perspective, drawing mainly on other descriptions (French, 1986a; 1986b; 1987). Some of these technologies have already been referred to in Chapters 8 and 9. Then, network equilibrium modeling concepts and capabilities are described, emphasizing their static, dynamic and finally quasi-dynamic assumptions in relation to the requirements for vehicle navigation. Finally, I speculate about the potential contributions of network equilibrium modeling to vehicle navigation and discuss the associated research implications.

10.2 ROUTE CHOICE CONSIDERATIONS OF CURRENT VEHICLE NAVIGATION TECHNOLOGY

Two European vehicle navigation systems have reached the stage of development of field testing: Autoguide and Ali-Scout. The Autoguide System, scheduled for field testing beginning in 1988 in London, may be described as a present-day realization of the ERGS concept. A central computer continually updates routes from intersections to destinations on the basis of current traffic conditions. This route information is stored in intersection-based, electronic signposts. As a vehicle approaches each intersection so equipped, an onboard computer transmits to the signpost, via inductive loops buried in the roadway, its destination, type of route, and preference of route type, if any. The signpost transmits back to the vehicle directions pertaining to that intersection (turn right, turn left, straight on), and this information is communicated to the driver visually and audibly.

The central computer accumulates data from the signposts and updates a data base of traffic conditions throughout the network by time of day and day of week. These data, together with the latest information on abnormal events, are used to determine best routes using standard minimum cost algorithms. The signposts are updated at regular intervals of perhaps 15 minutes, unless an incident requires more immediate action (Jeffrey et al., 1987). The computation of multiple routes is not considered in available descriptions.

The Autoguide System is reactive in the sense that it appears to rely on current traffic conditions, including traffic flows resulting from Autoguide instructions. Therefore, if these instructions result in delays, then traffic may be further rerouted away from these locations. About 750 signposted intersections are judged to be sufficient for a London system covering the entire area within the M25 circumferential motorway. On this basis the average distance between signposts would be two km ranging from one-half km in central London to five km at the periphery. Is an allocation of vehicles to multiple routes necessary at this density to avoid system-generated congestion? The answer will depend in part on the detail at which destinations are considered; it appears this issue could be productively studied using network models.

The Ali-Scout System, which is scheduled to undergo large-scale field testing in West Berlin beginning in 1988, combines aspects of the Autoguide System with an onboard map display of the recommended route. Onboard equipment determines the vehicle location by dead-reckoning and the location is displayed using map matching techniques. Recommended route data are received when vehicles pass infrared beacons, and transformed into graphic driving directions to the driver input destination. The equipped vehicle also transmits to each beacon it passes its travel history since passing the last beacon; thus, the equipped vehicles serve as their own traffic sensors (French, 1987). Available documentation does not discuss the issue of multiple routes and the possibility of system-generated traffic congestion by advising too many vehicles to use the same route.

A more advanced system under development by Philips called Carin combines the onboard features of Ali-Scout with storage of map data on compact disc and use of a car radio sub-carrier to piggyback traffic data on commercial FM broadcasts. The latter capability is presently available with Blaupunkt's ARI system in some parts of Europe. The system includes a route-search algorithm and provision of route information to the driver (French, 1987). Route information is described as being optimized for minimal travel time under prevailing traffic conditions; however, no discussion of the issue of multiple vs. single routes is provided.

Two Japanese systems are also under development. The CACS project already mentioned belongs to the same class as ERGS and Autoguide (Fujii, 1986; Yumoto et al., 1979). As a current report of its status is included in the previous chapter, no further comment is warranted. The Advanced Mobile Traffic Information and

Communication System (AMTICS) is a highly integrated traffic information and navigation system under development by the Japan National Police Agency (Nakashita et al., 1988). Traffic information gathered by police agencies will be processed and broadcast to autos for map display together with the vehicles location and map information stored on a compact disc. The system will display not only current position and route, but also traffic congestion information, regulations, road works and parking information. A pilot system is being implemented in Tokyo beginning in 1988. AMTICS, so far as can be determined from available descriptions, will not offer specific route information to the driver. Rather, it enhances drivers' route choice decisions by augmenting their available information base, as well as modifying traffic control signals.

10.3 NETWORK EQUILIBRIUM MODELING CONCEPTS AND CAPABILITIES

In this section, concepts of transportation network equilibrium modeling are introduced and evaluated with respect to their potential contribution to the route guidance features of vehicle navigation systems. The presentation is relatively nontechnical and emphasizes underlying assumptions, capabilities and limitations of these models, rather than their analytical features. Mathematical formulations are briefly presented in two appendices. The exposition begins with the traditional static model, briefly considers the prospects for large-scale dynamic models, and concludes with a discussion of a proposed quasi-dynamic formulation.

10.3.1 Static Network Equilibrium Models

The fundamental assumption of network equilibrium modeling is that each traveler seeks to minimize his or her cost of travel. Here cost is used in a general sense to include travel time, travel costs such as auto operating costs or transit fares, terminal costs such as parking fees and less tangible costs such as comfort and convenience, all converted to a common metric. This metric is sometimes referred to as *generalized cost*, but here I will refer to this disutility of travel as simply *cost*.

If the costs of alternative routes available to each traveler were known and fixed, each traveler would achieve the above behavioural objective by choosing the minimum cost route. Since urban travel costs are not fixed, but vary with flows of vehicles on *links*, the street and highway segments of which routes are comprised, the route choice problem is not nearly so simple. That the link travel times experienced by each traveler increase with increasing vehicle flow is hardly surprising to the experienced auto traveler.

The introduction into consideration of the link flows corresponding to the travel choices requires that the time period be considered as well. Now instead of abstractly examining route choices, one must consider the number of travelers per hour, or other convenient time period during which the *rate* of travel departures may reasonably be considered *constant*. Moreover, it is necessary to define this trip rate in terms of its origin and destination. Since one is concerned with a flow of trips, it is natural and practical to speak of small areas of origins and destinations, or zonal aggregates of individual trips into origin-destination (O-D) flows measured in trips per hour.

Now it is possible to state the network equivalent of our original assumption that each traveler chooses a route that minimizes his or her travel cost, assumed to be known precisely. An equivalent statement is:

1. all used routes from each origin to each destination have equal travel costs, and
2. no unused route has a lower cost than a used route.

A moment's reflection will convince one that these two statements are equivalent to the original assumption; if this were not true, some traveler could reduce his or her cost by changing routes. The network flow configuration described, if these statements are true for every O-D pair, is termed a *network equilibrium*. For a unique equilibrium to be achieved mathematically, the link travel costs must be increasing with increasing flow on each link and the O-D flow rates must be constant during the time period considered. If these conditions are met, then route travel choices exist that result in unique link flows and route costs, satisfying the above assumptions. In general, however, the route choices are not unique. The formulation of the static model is given in Appendix 1.

The conditions stated above clearly imply that *multiple* routes from origins to destinations are a feature of a network equilibrium. Indeed, an equilibrium cannot be achieved mathematically without some multiple routes, if link travel costs are flow dependent. To what extent multiple routes do occur in the solution of the static formulation of the problem is not revealed by the mathematical analysis. Likewise, in what ways the phenomenon of multiple routes depends on the size of the origin and destination zones, as measured both in geographical area and number of trips, is also unclear; however, it appears that as zones become smaller, fewer multiple routes are required. Moreover, the extent of multiple routes between O-D pairs in the mathematical formulation depends on the detail of the representation of the street network: detailed networks are likely to require fewer multiple routes than aggregated network representations.

Some insight into the answers to these questions can be obtained by examining the structure of algorithms for solving the network equilibrium model and the characteristics of the numerical solution. The algorithm for the static network equilibrium model solves a constrained minimization problem whose solution properties correspond to the above behavioural assumptions. The algorithm has been successfully applied to networks with tens of thousands of links and millions of O-D pairs. For example, in the Chicago region, highway networks with thirty thousand links and over three million O-D pairs are routinely solved to acceptable levels of precision (Eash et al., 1983).

The algorithm employed is iterative in nature. It solves the problem by temporarily fixing the link travel costs at the values corresponding to the link flows in the current solution. The minimum cost routes corresponding to these costs are determined; the total link flows that would result from placing each O-D flow on its minimum cost route are then obtained and averaged with those in the current solution, where the averaging weight is chosen to minimize a particular function of these flows. This process continues until the value of the function being minimized is sufficiently close to the unknown minimum.

The number of iterations performed determines the maximum number of possible routes between each O-D pair in the solution. The number of distinct routes is not explicitly determined since the used routes are normally not retained. For large congested networks, five to eight iterations of the algorithm are generally sufficient for planning purposes, implying that the number of used routes in the solution probably ranges from two to eight, except for trips between adjacent areas. However, some of the routes may be trivially different.

The model, algorithm and numerical results described above pertain to the case in which travel costs and shortest routes are assumed to be known. How the routes and costs are known by the travelers is not considered by this *deterministic* approach (Horowitz, 1984). In a generalization of this model, travelers are assumed to choose the route with the lowest *perceived* cost. Since these perceptions are subject to errors, then additional routes are likely to be used, as compared with the deterministic case. Indeed, in the solution to one version of this *stochastic* route choice model, some small portion of trips between each O-D pair are found to travel on every route, regardless of its excess cost above the minimum cost route. An unsolved issue with regard to this model is how to enumerate the routes in the solution algorithm. Available methods for finding all possible routes in a network are inadequate for large networks.

What are the implications of this static network equilibrium model for route guidance? First, advising all drivers of a single minimum cost route based on current flows is unlikely to be helpful. The initial solution to the network model is often found in this way; however, this solution is generally quite poor if the network is moderately to heavily congested. Therefore, even from this simple static model, it is clear that advising all drivers to take the same route from a given origin to destination is counterproductive. This is not true if the number of trips between each O-D pair is very small (e.g. one or two) and the routes advised for similar O-D pairs are somewhat different. I know of no studies of this effect; however, the areas of origins and destinations in the Chicago model are generally 2.6 or 10.4 sq. km., which is fairly small compared to the spacing of the signposts in the Autoguide System, which determine the extent to which information on different routes can be made available.

Second, as already noted, the static equilibrium model is intended to pertain to a relatively long time period as compared with trip travel times. In other words, a model of a 15 minute period with 45 minute trips is not meaningful. For this reason, these models are generally applied to one or two hour time periods. Since the trip departure rate on an O-D basis is unlikely to be constant for such a long period, the model is only a very rough approximation of reality. Since shorter time periods during which departure rates may be constant are unrealistic from a trip length point of view, it is clear that the static model is rather inappropriate for this application. Moreover, as discussed in the Introduction, departure time is an important choice dimension in vehicle navigation systems; to be able to consider this choice, a model incorporating it is necessary.

Third, the static model as usually implemented in practice does not consider delays resulting from conflicting flows at intersections. Instead, travel delays are assumed to occur between intersections based on traffic flow and on street characteristics such as width. If traffic signals are present, they are conventionally assumed to be set to some predetermined ratio of green time for each approach, such as 50/50. In recent years, more realistic network models incorporating conflicting flows at intersections have been devised, and the corresponding mathematical properties explored (Dafermos, 1982). Implementation of such models has progressed slowly; they are not yet used in practice. For these models to be potentially useful for vehicle navigation, however, it seems clear that the way in which intersections are modeled must be improved.

10.3.2 Dynamic Network Equilibrium Models

During the last few years, several research papers on dynamic models have appeared. As their name suggests, the time of day is explicitly represented in such models. For this reason, in these models the departure time, as well as route, becomes a principal choice variable. To represent traffic delays in such models, one may apply the technique of queuing theory, which describes the time dependency of travel delays in terms of the number of vehicles approaching the link and the link capacity.

The objective of such models is generally stated in terms of choice of departure time and route in order to reach a destination (e.g. a workplace) at a specified time at the minimum generalized cost including schedule delay, or the difference between the desired and actual time of arrival. Sometimes, a small slack time is permitted; for example, if a traveler arrives less than five minutes early, no schedule delay is incurred. But if the traveler arrives more than five minutes early, then this is regarded as a schedule delay. The objective of the model, then, is to choose the route and departure time for each user so as to minimize his/her total journey cost including schedule delay (Ben-Akiva et al., 1986.).

The principal difficulty with such models, as developed so far, is that they can only be solved for very small networks such as one O-D pair with two routes, one of which has a bottleneck where queuing occurs. Even these very simple models may not have a unique solution. Although such models may ultimately be useful, they do not seem to represent an approach which can form the basis of a vehicle navigation system in the near future.

10.3.3 Quasi-Dynamic Network Equilibrium Models

Inasmuch as truly dynamic models appear to be intractable, even for networks of modest size, a quasi-dynamic approach is suggested. A similar concept has been proposed by Hamerslag (1988); also van Vliet (1982) has suggested a heuristic approach to this problem. The concept of a quasi-dynamic model is one of a linked sequence of static submodels; to facilitate their description, it is assumed that travelers are sensitive only to variations in travel time in this model. Each submodel pertains to a fixed time period, t, of say 10 minutes duration. For each time period, fixed O-D flow rates prevail. This rate of flow occurs on the shortest travel time routes from each origin out to a time boundary of 10 minutes towards each destination during time period t. The location of the first time boundary depends upon the average delays encountered on links at the total prevailing link flow rates.

At the end of time period t, each O-D flow rate that began in time period t reappears as a new flow rate in time period t+1 from its location on the first time boundary to the final destination. During period t+1, the flow rates continue on their shortest time routes, as determined by total link flow rates in t+1, towards their destinations until a second time boundary at 20 minutes is reached. Concurrently, trips corresponding to O-D flow rates for period t+1 flow from each origin to a new first time boundary defined by travel conditions during period t+1. During period t+1, for example, some destinations of O-D flow rates originating in period t are reached. That is, the destination is within 20 minutes of the origin for a flow trip rate at period t. These flow rates are terminated. Flow rates in period t+1 which do not reach their destinations within 20 minutes terminate at the second time boundary, and continue from that point towards their destinations in time period t+2, and so forth. A formulation of the quasi-dynamic model is given in Appendix 2.

It appears that the quasi-dynamic model described in the above paragraphs can be solved by a generalization of the algorithm commonly used to solve the static model; however, to date this algorithm has not been implemented. If so, the solution of the model has the same multiple, used-route characteristic as the static model. Since O-D flows would be defined by time period of departure, as well as O-D pair, the number of used routes in this model is potentially much larger. The number of different used routes for each O-D pair and time period, however, might be smaller than in the static model, since in the quasi-dynamic model trips are disaggregated by time of departure. If somewhat different routes are determined by the algorithm for an O-D trip rate for time periods 1, 2 and 3, then this corresponds to three different routes in the static model. Such a distinction could be very important for a vehicle navigation system in advising drivers concerning best routes at a specific departure time interval.

Another application of the quasi-dynamic model concerns the prediction of departure times by O-D pair. Suppose the destination arrival rates are known by time period together with the O-D rates for all time periods of the morning journey-to-work. What is the departure rate by O-D pair such that the O-D travel time plus schedule delay at the destination is a minimum? It appears that these O-D departure rates can be determined by solving the quasi-dynamic model with fixed arrival rates. This model has the potential to advise individuals concerning travel times corresponding to various departure times.

10.4 TOWARDS A REAL-TIME ROUTE GUIDANCE MODEL

This review of current and prospective network equilibrium modeling capability provides a basis for an initial description of the operation of a real-time route guidance system. By a real-time system, I mean one which is utilized to advise drivers about route choices during a peak congestion period based on current and prospective conditions. How would it work?

In order to advise drivers concerning route choices, a detailed, quasi-dynamic network equilibrium model of the peak period is required. The model is initially solved before the peak period based on historical data such as from the previous days and expected demands for the present weekday. As the peak period develops, observed flow rates replace predicted flows and the model solution for subsequent time periods is updated and disseminated to drivers.

Drivers with onboard navigators requesting route information are provided with minimum cost routes according to the requested criteria, such as value of time, prior route preferences and desired arrival time. Based on these criteria, route information is transmitted to the vehicle navigator and updated as conditions warrant.

In the event that many drivers request route information for common O-D pairs, the effects of providing the same route information to all drivers is evaluated. If a route saturation level is passed, drivers are advised to use alternative, but equal cost routes. Experience with this phenomenon, together with predictions based on the model, should be evaluated in order to determine whether route information provided has unduly added to the congestion level.

As the peak period progresses, disruptions in traffic flow caused by accidents, vehicle breakdowns, out-of-service traffic signals and other causes are considered. The model inputs are revised and the route solutions updated and disseminated as needed. Following the completion of the peak period, the flow data are routinely analyzed and historical files updated in preparation for subsequent days.

10.5 A RESEARCH AGENDA FOR REAL-TIME ROUTE GUIDANCE MODELS

Ideas set forth in the above sections are collected in this section in the form of a research program on the extension and application of network equilibrium models for vehicle navigation. The research agenda is largely concerned with three questions. First, what is the extent and character of the incidence of multiple routes between O-D pairs in solutions of the network equilibrium model in heavily congested urban networks? Second, can network route choices and link flows be adequately described and predicted by the quasi-dynamic formulation described? Third, if the quasi-dynamic model can be implemented, is it an adequate basis for the route guidance component of a vehicle navigation system?

The first question concerning the incidence of multiple routes is relatively straightforward to answer. Routes between each O-D pair which is used in the equilibrium solution for a large network can be readily stored and compared to determine if they are exactly alike. The number of different routes should be tabulated by length of the route, since short routes may have less possibility of being different. Then, the incidence of multiple routes should be studied as a function of the congestion level. Highly congested networks will surely have more multiple routes than lightly congested ones. Finally, the size of the origin and destination zones should be varied. Larger zones might be expected to result in more multiple routes than smaller ones.

Determination of whether routes are exactly coincident is easy. Determination of whether they are substantially different is more difficult, especially if performed for millions of O-D pairs in a large zone system. Some criteria for judging whether routes are different include the following: Is the proportion of links that are common to two routes relatively low? Does one route include a freeway segment whereas the other uses arterial streets entirely? Is one route considerably longer than the other in terms of over-the-road distance? If the answers to one of these questions is yes, then the routes would seem to be different.

This question of multiple routes between O-D pairs also raises an issue pertaining to the solution algorithm. The algorithm commonly employed (LeBlanc et al., 1974) solves the problem in terms of link flows; the routes themselves are discarded once the link flows are determined. A competing algorithm (Dafermos and Sparrow, 1969; Leventhal et al., 1973) solves the problem in terms of route flows by shifting flows from the highest cost route to the lowest cost one for each O-D pair. This algorithm would seem to be more suitable for addressing this issue since one could then determine what proportion of the flow was on the first, second, third,.. routes ranked in order of flow from heaviest to lightest. This route-based algorithm has not been widely used in the past because of its large memory requirements; however, the algorithm is claimed to converge more quickly than its link-based counterpart. Thus a by-product of using the route-based algorithm would be an opportunity to evaluate its performance on the current generation of supercomputers.

The second question concerning quasi-dynamic models also raises a number of important research issues similar to the ones just discussed pertaining to the level of network congestion, zone size and the similarity in routes. Another question concerns the appropriate length of the departure time interval and the width between time boundaries; both were taken to be 10 minutes in the above description, but they need not be the same. Indeed, the second should be measured in terms of generalized cost.

A departure time interval of 10 or 15 minutes seems correct because this is the smallest time interval that travelers seem to be able to distinguish in reporting their departure times in surveys. The width between time or cost boundaries is more related to the detail of the network. One would like the time or cost of an individual link to be small compared to this width, say not more than 20%, so that entire links mainly lie between boundaries. Not discussed above is the need to partition flows and costs for links lying across these boundaries.

In the final analysis, the answers to these questions depend upon the ability of the quasi-dynamic model to describe more adequately the variations in link flows and route flows during the peak period. Although data are generally available on selected link flows by short time intervals, it is uncommon for O-D tables to be tabulated by departure time intervals as detailed as needed for this model. Until such data can be collected, it may be necessary to solve the model for specified arrival times and evaluate the predicted departure rates by origin with the observed rates. Such a comparison may be possible using 1990 Census data for a large metropolitan area such as Chicago.

To examine the third question concerning the adequacy of the model for a vehicle navigation system, it may be necessary to consider the larger question of the benefits of such systems. To determine these benefits, a model of route choice based on incomplete knowledge is needed. A dispersed route choice model based on assumed errors of perception could represent this condition. Then, selected travelers could be provided with information on best routes and the reduction in travel times and costs determined. As a lower limit, the cost minimizing model of route choice could be compared with the dispersed route choice model. An important question in this regard concerns how dispersed is route choice without vehicle navigation. Almost nothing is known about this question since data on route choices for trips with similar O-D pairs is almost never collected. A small sample survey of this question could be very helpful in understanding this phenomenon.

In conclusion, there are many unanswered questions about route choice models in spite of ongoing research activities over the last 20 years. Although much progress has been made, the research was typically directed at understanding overall model properties and solution algorithm performance, rather than specific questions important for an application such as vehicle navigation. A research program focussed on these issues could be very helpful in evaluating the feasibility and promise of vehicle navigation systems.

APPENDIX 1: **THE STATIC TRANSPORTATION NETWORK EQUILIBRIUM MODEL**

Notation:

T_{ij}	=	flow rate from zone i to zone j in vehicles per time period
h_{ijr}	=	flow rate on route r from zone i to zone j in vehicles per time period
C_{ijr}	=	cost of travel per vehicle on route r from zone i to zone j
f_l	=	flow rate on link l in vehicles per time period
$c_l(f_l)$	=	cost of travel per vehicle on link l with flow rate f_l
$a_{l,ijr}$	=	1, if link l belongs to route r from i to j; 0, otherwise
n	=	number of zones of origin and destination
R_{ij}	=	number of routes from origin zone i to destination zone j

Definitions:

$f_l = \sum_i \sum_j \sum_r h_{ijr} \, a_{l,ijr}$ (link flow equals sum of route flows using the link)

$C_{ijr} = \sum_l c_l(f_l) \, a_{l,ijr}$ (route cost equals the sum of the costs of the links comprising the route)

The equilibrium link flows are determined by solving the following convex programming problem:

$$\min_{(f_l)} \sum_l \int_0^{f_l} c_l(x)\,dx$$

s.t. $\sum_r h_{ijr} = T_{ij}, \quad i,j = 1, \ldots n.$

$h_{ijr} \geq 0 \quad , \quad r = 1, \ldots R_{ij} \; ; \quad i, j = 1, \ldots n.$

The solution to the above problem is characterized by the following equilibrium conditions:

1. if $h_{ijr} > 0$, then $C_{ijr} = u_{ij} =$ equilibrium travel cost from zone i to zone j.

2. if $h_{ijr'} = 0$, then $C_{ijr'} \geq u_{ij}$;

3. if $C_{ijr''} > u_{ij}$, then $h_{ijr''} = 0$.

These statements are equivalent to the conditions stated on p. 8, and are commonly known as the Wardrop (1952) conditions.

APPENDIX 2: QUASI-DYNAMIC TRANSPORTATION NETWORK EQUILIBRIUM MODEL

Notation:

$t = 1, \ldots \tau$ = a short time period of 10 minutes length; e.g. 7:00 to 7:10

s = a short travel time interval from the trip origin; by convention, the firstinterval of 0 to 10 minutes is designated 0, the second interval of 10 to 20 minutes is designated 1, etc.

$a^{t,s}_{l,ijr}$ = 1, if link l belongs to route r from zone i to zone j in time period t and lies within time interval s.

= 0, otherwise.

All other notation is the same as in Appendix 1 except for the addition of a superscript t.

Definitions:

$$f^t_l = \sum_i \sum_j \sum_r \left(h^1_{ijr} \; a^{1,t-1}_{l,ijr} + h^2_{ijr} \; a^{2,t-2}_{l,ijr} + \ldots + h^t_{ijr} \; a^{t,0}_{l,ijr} \right),\; t = 1, \ldots \tau.$$

$$C^t_{ijr} = \sum_l \left(c^t_l (f^t_l) \, a^{t,0}_{l,ijr} + c^{t+1}_l (f^{t+1}_l) \;\; c^{t+2}_l (f^{t+2}_l) \, a^{t,2}_{l,ijr} + \ldots \right)$$

The equilibrium link flows are determined by the solution to the following convex programming problem:

$$\underset{(f^1_l, \ldots f^T_l)}{\text{minimize}} \quad \sum_{t=1}^{T} \sum_l \int_0^{f^t_l} c^t_l(x)\, dx$$

$$\text{s.t.} \quad \sum_i \sum_j \sum_r h^t_{ijr} = T^t_{ij},\; t = 1, \ldots \tau; \;\; i, j = 1, \ldots n$$

$$h^t_{ijr} \geq 0,\; r = 1, \ldots R_{ij};\;\; i,j = 1, \ldots n$$

The solution is characterized by the following conditions:

1. if $h^t_{ijr} > 0$, then $C^t_{ijr} = u^t_{ij}$ = equilibrium travel cost from i to j for trips leaving i during period t

2. if $h^t_{ijr'} = 0$, then $C^t_{ijr'} \geq u^t_{ij}$

3. if $C^t_{ijr''} > u^t_{ij}$, then $h^t_{ijr''} = 0$.

REFERENCES

Beckmann, M., McGuire, C. B. and Winsten, C. B. (1956) *Studies in the Economics of Transportation*, New Haven, Connecticut: Yale University Press.

Ben-Akiva, M., dePalma, A. and Kanaroglou, P. (1986) "Dynamic model of peak period traffic congestion with elastic arrival rates", *Transportation Science 20:* 164-181.

Boyce, D. E. (1988) "Route guidance systems for improving urban travel and location choices", *Transportation Research 22A:* 275-281.

Dafermos, S. (1982) "Relaxation algorithms for the general asymmetric traffic equilibrium problem", *Transportation Science 16:* 231-240.

Dafermos, S. and Sparrow, F. T. (1969) "The traffic assignment problem for a general network", *Journal of Research of the National Bureau of Standards 72B*: 91-118.

Dial, R. (1987) "The Etak digital map database: potential applications in highway transportation", *Technology Options for Highway Transportation Operations*, Institute of Transportation Studies, University of California, Berkeley, California.

Eash, R. W., Boyce, D. E. and Janson, B. N. (1979) "Equilibrium Trip Assignment: Advantages and Implications for Practice", *Transportation Research Record 728*: 1-8.

Eash, R. W., Chon, K. S., Lee, Y. J. and Boyce, D. E. (1983) "Equilibrium Traffic Assignment on an Aggregated Highway Network for Sketch Planning". *Transportation Research Record 994:* 30-37.

French, R. L. (1986a) "Historical overview of automobile navigation technology", paper prepared for the IEEE Vehicular Technology Conference, Dallas, Texas.

French, R. L. (1986b) "Automobile navigation: Where is it going?" paper prepared for the IEEE Position, Location and Navigation Symposium, Las Vegas, Nevada.

French, R. L. (1987) "The evolving roles of vehicular navigation", *Navigation: Journal of the Institute of Navigation 34*: 212-228.

Fujii, H. (1986) "The CACS project and now: dynamic route guidance as the final target", paper presented to the 65th Annual Meeting of the Transportation Research Board, Washington, D.C.

Hamerslag, R. (1988) "A three dimensional assignment in time-space", paper presented at the University Transport Study Group Annual Conference, London.

Horowitz, J. L. (1984) "The stability of stochastic equilibrium in a two-link transportation network", *Transportation Research 18B:* 13-28.

Jeffrey, D. J., Russam, K. and Robertson, D. I. (1987) "Electronic route guidance by Autoguide", presented at the 15th Annual Meeting, Planning and Transport Research and Computation Co., University of Bath, England.

LeBlanc, L. J., Morlok, E. K. and Pierskalla, W. P. (1974) "An efficient approach to solving the road network equilibrium traffic assignment problem", *Transportation Research 9:* 309-318.

Leventhal, T., Nemhauser, G., and Trotter, Jr., L. (1973) "A column generation algorithm for optimal traffic assignment", *Transportation Science 7:* 168-176.

Nashashita, H., Okamoto, H. and Kawabata, T. (1988) "Advanced mobile traffic information and communication system", presented at the 67th Annual Meeting of the Transportation Research Board, Washington, D. C.

Rosen, D. A., Mammano F. J. and Favout, R. (1970) "An electronic route-guidance system for highway vehicles", *IEEE Transactions on Vehicular Technology VT-19:* 143-155.

van Vliet, D. (1982) "SATURN - A modern assignment model", *Traffic Engineering and Control 23*: 578-581.

Wardrop, J. G. (1952) "Some theoretical aspects of road traffic research", *Proceedings of the Institution of Civil Engineers, Part II, 1*: 325-378.

Yumoto, N., Ihara, H., Tabe, T. and Naniwanda, M. (1979) "Outline of the CACS pilot test system", paper presented at the 58th Annual Meeting of the Transportation Research Board, Washington, D. C.

THE FUTURE OF SURFACE TRANSPORT

(b) Goods Transportation

Chapter 11

THE FUTURE OF TRUCK TRANSPORTATION IN EUROPE

Hellmuth St. Seidenfus
University of Münster, Federal Republic of Germany

11.1 ECONOMIC EUROPE TO THE YEAR 2000

11.1.1 Hypothesis

Long term forecasts on the economic development of the EEC and EFTA member states are generally optimistic despite the continuing problems affecting world-wide economic growth such as unsatisfactory structural adjustments, unemployment and imbalances in foreign trade.

11.1.2 Reasons

The establishment of the European Common Market in the nineties will generate substantial economic impulses which will lead to average annual growth rates of GNP of 3 to 4 percent in the region. Growth factors are especially:

* *Increasing returns to scale in production and marketing.*
  Concentration and mergers will continue. Larger firms are more capable of satisfying the demands of a larger market as long as the "industrial targeting" of Japan and the Newly Industrializing Countries of East Asia is curbed.
* *Improved utilization of specialization.*
  The larger market will increase the possibilities of making use of the division of labour within the Common Market as long as the mobility of productive factors, goods and services is guaranteed and non-tariff barriers are abolished.
* *Increasing real-term income.*
  The increase of real-term income due to improved productivity will increase supply and demand of goods and services. This will result in a higher

standard of living within the market of 320 million people as long as reasonable wage policies are agreed on and unemployment is reduced.

* *Increased innovation.*

  Intensified competition will compel firms to put more effort into product and process innovation. A prerequisite is that investments and private consumption develop in a dynamic equilibrium. Firms will only be willing to invest if satisfactory returns are expected and the long-term economic data supplied is encouraging.

A further source of growth can be seen in the fact that the weaker member states and regions of the Common Market will be helped in their economic development by coordinated support from the other member states. This policy of economic support is based on Art. 130 b of the Treaty.

Prerequisites are:

* spillovers from the member states where economic growth is in progress;
* economic policies oriented towards real-term growth potentials in the developing member states;
* appropriate use of the instruments of the Common Market in the field of structural development and financial aid.

For the past few years the EFTA member states have been seeking closer relationships with the member states of the EEC. Switzerland and Austria for example have been indicating their willingness to join the EEC. Every new accession will immediately improve the growth potential within the newly established market.

11.1.3 Result

The reasoning has so far indicated that there are still major problems to be solved and that it is by no means certain that they will be overcome. Therefore the following forecasts can only by regarded as trends which will be exemplified by means of scenarios - the prerequisites of which will not be discussed in detail (Keimel et al., 1981).

The most important growth factors in a technology scenario are a new wave of technological innovations and a substantial increase in world-wide trade on the basis of an annual growth of population of 0.23 percent (1980-2000). Technological innovation will ensure that no barriers to growth will occur either due to lack of resources or due to environmental restrictions. The economic growth as forecast for some of the developing countries, but especially the further increase of intra-European trade, are immediately responsible for the impulses of world-wide trade.

Accordingly, the following annual growth rates for the EEC member states are forecast for the years 1990 to 2000 (in percent):

GDP	4.5	Technology scenario
exports	8.8	
employment	0.37	

The figures are higher than in the years 1980 to 1990 and will only be reached if economic growth is accepted as one of the major government policies.

A public service scenario treats quantitative growth as inferior to other goals which are generally described as quality of life. In addition, economic activities are shifted into the tertiary industries, one of which is transport. On the one hand, the structural changes experienced in this context are a result of consumption habits. The demand for goods of the primary industry decreases in relative terms and the demand for durable consumer goods and services increases in absolute terms. On the other hand, the sectoral changes expected are due to increasing complementary relationships between industrial growth and the need for goods and services in the tertiary industry. Highly specialized workers need specific teaching methods; complicated technical, commercial and financial services give consulting engineering firms, trade firms and banks a more important role; growing mobility leads to a higher demand for transport.

For the years 1990 to 2000 this scenario yields the following results:

GDP	2.3	Public service scenario
exports	4.0	
employment	0.17	

Both scenarios try to describe plausible future developments.

The results endorse the optimistic forecasts as far as general economic growth is concerned. Eventhough the transport sector as part of the tertiary industry will prosper as far as the scenarios are concerned, a stringent forecast of the future of the European road transport markets cannot be made. To do so it is necessary to disaggregate the global figures in order to isolate the elements relevant for road transport.

11.2 CONTINUING STRUCTURAL CHANGES

11.2.1 Hypothesis

Economic growth is always accompanied by structural changes. These structural changes have always forced the road transport markets to offer more sophisticated services than the other continental modes. This development will continue even more emphatically in the future.

11.2.2 Reasons

The noticable intra-sectoral and inter-sectoral structural changes cannot be associated with just one reason but are due to the following developments:

* different increases in demand for goods and services in the various sectors.
  Since income elasticities which are responsible for the demand of goods and services differ from one another and vary in time when subject to saturation or changes in consumer behaviour, real-term increases in income will lead to structural changes.
* different increases of demand and supply of factors of production depending on the sectors dealt with.
  Factors of production are often subject to natural barriers limiting their expansion. On the other hand substitutions between the factor labour and the factor capital occur when the ratios between the factor prices change. The attempt to evenly balance the marginal productivities of the utilized factors will necessarily lead to variations of factor demands.
* different realization of engineering progress depending on the sectors dealt with.
  The sectors all have different propensities to innovation. Engineering progress leads to different degrees of savings of factors and therefore also to variations of demand for factors.

There are several hypotheses for these differences in structural changes, providing some hints as to the intensity and dynamics of the developments which are to be expected:

* *Demand cycles.*
  This thesis is based on the assumption that changes in demand for specific goods lead to adaptations in production that immediately affect product and labour ratios. The variations are due to demographic factors, the increase of real income and the allocation to income classes, saturation and modification of personal values. The changing of demand levels, demand structures and the variation of demand as time passes are responsible for structural modifications between the consumer industry and the capital-goods industry.
* *Product cycles.*
  This thesis is based on the assumption that modifications of product structures are due to the moving of production localities. High standard mass production is moved to low-wage countries (developing countries). On the other hand product cycles change when the import levels of high productivity technology decrease and the possibilities of imitating engineering processes decrease. New engineering procedures will have to be invented on their own. This is why sectoral changes occur between the secondary and tertiary industries.

* *Over-industrialization.*
  In highly developed countries discrepancies can occur between the supply and demand sides of tertiary industry in comparison to secondary industry. Adaptation potentials can deteriorate which means that production levels between the sectors will have to change.

It seems that all three hypotheses are able to explain the structural changes of the past. It must be expected that these changes will more or less continue up to the year 2000.

11.2.3 Result

These general economic structural changes have affected the transport sector both as far as its level and its structure are concerned. The level effects have to do with the fact that an increase in GDP will lead to an increase in demand for transport services. The coefficients of elasticity of rail, road and inland waterway transport in the EEC member states in the sixties and seventies were 0.8 for domestic transport and 1.3 for international transport. In respect to the development of GDP in the years 1970 to 1982 the elasticities were as follows (OECD 1986, p. 44):

Austria	1.17
Denmark	1.11
France	0.57
West Germany	0.78
Italy	1.91
Netherlands	0.58
Norway	0.78
Spain	1.35
UK	1.23

The substantial differences between the figures are due to the different values of transported goods in these economies. "This phenomenon has a fairly logical explanation. Output is expressed in, say, francs and traffic in tonnes-km: in other words the elasticities are in large measure bound up with the value added/tonnes-km ratio of the goods carried. This ratio increases as goods are processed. ... The result is a lower elasticity of transport with respect to production since, for a high value in output terms, traffic in tonnage terms will expand only marginally as a percentage" (OECD 1986, p. 45/46).

Accordingly, countries with a low GDP per capita will have elasticities higher than 1; West European countries, however, will have values below 1. For the EEC member states this will result in elasticity coefficients of between 0.60 and 0.65 for domestic traffic and 1.1 to 1.2 for international traffic. This will mean annual growth rates of 2.0 to 3.5 percent for international transport volumes as well as increased average distances (Keimel et al. 1981, p. 137).

These level effects give rise to the assumption that transport levels in Western Europe will increase further. Different developments are taking place in the various sectors of production. The production structures in the various countries will change. These freight structure effects will affect the modes in different ways. Depending on their technical, economic and organisational qualities they will adapt to the changing demand differently.

Table 11.1 Changes in Average Annual Market Shares in tonnes-km, Percentages (1970-1982)

	Road	Rail	Waterway	Coastal
Austria	+ 1.97	- 1.79	- 0.18	.
Belgium	+ 0.93	- 0.33	- 0.60	.
Denmark	+ 0.56	- 0.56	.	
Finland	+ 0.80	- 0.10		- 2.40
France	+ 0.94	- 0.68	- 0.26	.
Germany	+ 1.22	- 1.02	- 0.21	.
Italy	+ 7.60	- 0.30	- 2.10	+ 1.05
Japan	+ 0.43	- 0.88	.	+ 0.45
Luxembourg	+ 2.41	- 2.26	- 0.15	.
Netherlands	+ 0.5	- 0.24	- 0.30	.
Norway	+ 0.53	- 0.53	.	.
Spain	+ 0.55	- 0.55	.	.
Sweden	+ 0.10	- 2.40		+ 4.70
Switzerland	+ 0.54	- 0.46	- 0.03	.
United Kingdom	+ 0.61	- 0.61	.	.
United States	+ 0.25	- 0.33	+ 0.07	.
Overall trend	+ 1.25	- 0.80	- 0.42	.

The freight structure effects are a result of the following phenomena:

* increasing share of advanced investment goods in the overall industrial production ("intelligent products") and decline of primary goods;
* intensified international trade within the Common Market;

* miniaturization of products;
* organisational changes in the purchase and distribution structures of producers and retailers.

The impact of this increased demand for transport activities will be enhanced by substitution effects. These occur between the various modes of transport due to the different technical, economic and organisational developments in those modes which have modified their competition potentials (see Chapters 5 and 6).

In the years 1970 to 1982 the level, structure and substitution effects have influenced the modal split between the domestic modes in more or less the same manner (OECD 1986, p. 49).

It is apparent that the developments were in favour of road transport. As shall be shown in the following chapters, an analysis of the modified requirements of the consignors and the possibilities of the modes to meet these requirements allows us to predict the future of European transport markets.

11.3 NEW REQUIREMENTS IN THE TRANSPORT SECTOR

11.3.1 Hypothesis

The development of production structures, a continuing diversification of products and a more intense competition in transport markets will lead to a decline of the importance of transport prices in rapidly growing economic sectors. Accordingly, quality factors such as speed, reliability and adaptability will become more important when choosing modes. The decentralization of production localities is a further result of an increasing division of labour that requires new solutions in the field of logistics.

11.3.2 Reasons

The producers and retailers are confronted with new market situations. Many seller's markets have developed into buyer's markets. Intensified competition forces them to implement a strategy of product differentiation to remain competitive. Product life cycles are becoming shorter and the risks of producing for an anonymous market are increasing.

Competition in buyer's markets will lead to price reductions. That is why there is an urge to reduce the amount of fixed capital by cutting down stocks both on the purchasing and selling side of firms. The aim in this case is to increase productivity in areas other than production. Apart from the quality of the products, a substantial level of service qualities will be needed. Services in this sense mean:

* constant delivery capability
* short delivery times
* ultimate reliability
* low risk of damaging fragile goods
* efficient flow of information.

At first, the improvement of service levels increased overall costs. In the central, intermediate and exterior stores the amount of products increased. Due to their high quality and the value they represented, a large amount of money was tied up in stock. In addition, the increased number of smaller goods (miniaturization) needing shorter transport times as well as an intensified international trade led to an increase in organisational expenditure.

Industry reacted in two ways:

* just-in-time deliveries and just-in-time production are gaining in importance as can be seen in the car and electro-mechanical industries
* the number and proximity of stores were reduced. In West Germany, for example, 20 to 30 stores were required for nation-wide distribution services in the sixties. In the seventies the number was reduced to about 10. Currently, attempts are being made to cut that number to 2 or 3.

The experiences of American firms in introducing Japanese concepts of stockless production management (Kanban), or other successful attempts in Europe, show that it is possible to reduce these costs by means of logistical systems. The potentials of logistical systems are

* *That the level of capital tied up in stock is reduced.*
  More frequent, smaller and more precise deliveries reduce stocks. In order to achieve reasonable transport lots goods have to be batched. Integrating the transport necessities of several consignors seems a suitable policy.
* *That service quality is improved according to competitive pressures.*
  The integration of logistic sub-systems into a single logistic system will guarantee a consistent and stringent quality of logistic services. This is important for developing corporate identities.
* *That organisational expenses are reduced.*
  Systems planned in anticipation of possible problems are able to replace highly expensive, uncoordinated ad hoc decision taking. The organisation of information flows helps avoid malfunctions.

Table 11.2 Factors Affecting the Importance of Road Transport for Overall Economic Growth

<table>
<tr><th colspan="2" rowspan="2">Product category[a)]
NST/R - CSTE</th><th rowspan="2">Proportion of product categories in foreign trade in 1981 prices</th><th colspan="2">Proportion of road transport in international trade (in tonnes)</th><th rowspan="2">Growth rates of international road transport per 1 % increase of international trade</th><th rowspan="2">Variation of volume increases of international road transport per 1 % variation of road transport prices in relationship to those of rail and waterways</th></tr>
<tr><th>min. (*)</th><th>max.(*)</th></tr>
<tr><td>NST/R 0+1</td><td>Agricultural products live animals foodstuffs and fodder</td><td>12 %</td><td>58 %</td><td>69 %</td><td>0,786 %</td><td>0,825 %</td></tr>
<tr><td>NST/R 2</td><td>Solid mineral fuels</td><td rowspan="2">16 %</td><td>8 %</td><td>10 %</td><td>0,560 %</td><td>(**)</td></tr>
<tr><td>NST/R 3</td><td>Non solid fuels and petroleum products</td><td>9 %</td><td>12 %</td><td>0,563 %</td><td>0,076 %</td></tr>
<tr><td>NST/R 4</td><td>Iron ores and metalwaste</td><td>1 %</td><td>4 %</td><td>6 %</td><td>0,471 %</td><td>2,657 %</td></tr>
<tr><td>NST/R 5</td><td>Metal products</td><td>7 %</td><td>40 %</td><td>47 %</td><td>0,774 %</td><td>0,047 %</td></tr>
<tr><td>NST/R 6</td><td>Crude and manufactured minerals, building materials</td><td>5 %</td><td>35 %</td><td>42 %</td><td>0,508 %</td><td>0,075 %</td></tr>
<tr><td>NST/R 7</td><td>Fertilizers</td><td>4 %</td><td>20 %</td><td>30 %</td><td>0,575 %</td><td>2,655 %</td></tr>
<tr><td>NST/R 8</td><td>Chemicals</td><td>11 %</td><td>62 %</td><td>75 %</td><td>0,900 %</td><td>0,007(***)</td></tr>
<tr><td>CSTE 17</td><td>Metal products</td><td>13 %</td><td rowspan="4">72 %</td><td rowspan="4">80 %</td><td>0,857 %</td><td>0,096 %</td></tr>
<tr><td>CSTE 18</td><td>Machines and transport equipment</td><td>19 %</td><td>0,760 %</td><td>0,095 %</td></tr>
<tr><td>CSTE 19</td><td>Other manufactured articles</td><td rowspan="2">12 %</td><td>0,944 %</td><td>0,094 %</td></tr>
<tr><td>NST/R 9</td><td>Manufactured articles not mentioned elsewhere</td><td>0,906 %</td><td>0,096 %</td></tr>
</table>

a) NST/R - CSTE: The standard goods classification for transport statistics, abbreviated as NST (Nomenclature uniforme de marchandises pour les Statistiques de Transport), came into use following a recommendation in 1961 by the Commission of the European Communities.
The NST was prepared in close cooperation with experts from the Member States of the European Communities, the United Nations Economic Commission for Europe in Geneva (ECE/UNO), and also other international bodies. There is complete correspondence between it and the Commodity Classification for Transport Statistics in Europe (CSTE) prepared by the ECE/UNO, except for six basic headings where a further breakdown was required to meet specific Community requirements.

(*) calculated by IRU on the basis of available statistics

(**) insignificant results

(***) figures only valid for waterways

The aim to reduce stores and transport activities, as well as the attempt to increase the quality of managerial purchasing and distribution, have heightened the interest in establishing a sophisticated logistical management.

The progress achieved in the division of labour in Western Europe and the subsequent success in decentralizing production will lead to new production areas in the Mediterranean region. Accordingly, the linking of the EFTA member states to the EEC will increase the tendencies of decentralization. These new centres of economic activity will seek closer connections to central Europe and in doing so will intensify the needs for new logistical systems. Both transport distances and the proportion of international transport will increase.

Therefore, the transport volume (measured in tonnes) will not keep pace with the tonne-km transport as well as with the development of GDP. The smaller overall weight of transported goods and their increased value will result in transport prices becoming less important whereas the quality of transport procedures such as on time deliveries, frequencies, and rapid handling of goods will become more and more decisive.

That is not to say that transport prices will not be relevant, but simply that quality standards will become more important. A list of the factors that demonstrate how important road transport is for economic development in Western Europe is shown in Table 11.2 (IRU 1984, P. 11). At the same time it also becomes obvious that transport prices are losing relevance.

The table shows that a variation of relative prices between ground transport modes will have no significant effect on modal split. The only goods that show any sensitivity to price variations are bulky ones, which is of no great relevance for road transport. Those goods which play a major role in road transport in the OECD countries are hardly affected by price variations - the effets being in the region of 1 percent or less. Deregulation of the European transport markets, and the decline of transport prices, will indeed improve the competitiveness of road transport.

11.3.3 Result

The impact of structural changes in industrial production in the European member states has influenced requirements in the transport sector. "The structure of production is dispersing geographically and increasing subcontracting and adding more refining phases after current processing lines, which will make the flows dispersed and thinner. The lower inventories require more accurate time of arrival, and the more refined products are more fragile" (OECD 1986 p. 95). The manufacturing industries are inclined to replace production and warehousing costs by transport costs in order to improve productivity. For the transport sector the following tendencies are relevant:

* smaller units per trip
* faster and more reliable deliveries
* more precise timing
* more flexible dispositioning

The experiences in the past have shown that road transport is extremely capable of adjusting itself to geographical and temporal process changes in production. Rail transport, however, has tried to adjust by introducing faster trains and guaranteed delivery times, but has never come near to the flexibility of road transport. The unsatisfactory results in organising international freight transport by means of international combined transport is a further example of a lack of flexibility.

It seems plausible that the future needs of the consignors will urge the road transport sector to find further solutions in the years to come.

11.4 ADAPTION STRATEGIES IN THE ROAD TRANSPORT SECTOR

11.4.1 Hypothesis

In developing new logistical systems by dividing markets into segments and concentrating production units, transport firms are able to meet the consignors' requirements.

11.4.2 Reasons

11.4.2.1 *Developing overall integrated logistical concepts*

The problems of storing, transporting and the commissioning of goods as functions integrated into modern logistical concepts are nothing new in comparison to the problems the freight forwarders have had to solve in the past. What is new is that these individual tasks now have to be integrated into an entire logistical package. This generalizing strategy results in the freight forwarder becoming a distributor who is capable of meeting the individual requirements of the consignors.

If the freight forwarders are too inflexible, consignors develop their own customized logistical systems and use own - account traffic to implement the system. But even then there is a possibility for the freight forwarder to take part in the consignor's logistical system if he is able to propose improvements or if he is more efficient in any specific task. Any participation will mean establishing a form of vertical cooperation with the consignors and investing in nonmaterial intermediate inputs and material investments. Nonmaterial intermediate inputs have to do with the development of logistical concepts as well as the information systems needed for their correct functioning. Material investments are investments in central buildings for the batching, distribution, transshipment and handling of goods, as well as the installation of information and documentation centres.

Another task is to overcome the problem of management deficits - a question which seems to be more critical for freight forwarders than for consignors. Due to the expenses involved, it is difficult for small and middle-sized forwarders to introduce strategic management concepts which are assisted by modern management techniques. The economic structures of the EEC member states have also complicated the situation in that market impulses were suppressed. "The government price structures have made cost price calculation as a means of determining price floors obsolete. Fixed prices suppressed any urge to establish a system for estimating target costs, mistakes in investment calculations (e.g. in assessing optimal replacement dates for vehicles) were not penalized, firms were allowed to postpone the introduction of data processing without any serious consequences, substantial innovation of processes and products was not needed and management strategies were unnecessary because there were no long-term problems which needed to be anticipated" (Bretzke 1987, p. 62/63).

11.4.2.2 *The division of markets into segments*

If a freight forwarder is not able to establish general management strategies he has to take his chances in appropriate market segments that are not catered for by general freight forwarders. The strategy of specializing in certain markets is underlined by the fact that due to increasing geographical and factual diversifications of demand, the consignors are inclined to seek new logistical solutions.

Fields of specialization are

* junctions in areas such as sea ports;
* refrigerated goods, high speed transports (UPS), glass transports etc.;
* regions such asthe Far East;
* clients such as specific consignors.

It seems plausible to presume that road transport firms that do not succeed in establishing either a general or a specialized logistical concept will be forced to become sub-contractors to larger commercial freight forwarders; own account transport firms will have to merge with other companies to become competitive.

11.4.2.3 *Concentration by means of mergers and co-operation*

In order to achieve efficient solutions to the problems posed by the consignors, small and middle-sized firms will have to amalgamate or cooperate with other firms. The "drive for all-out informatics" is typical for tendencies of this kind. The OECD (1986, p. 131) quite correctly mentions that "... the industry must learn to offer a more sophisticated range of services in the framework of much larger organisation structures; it must transcend its fragmentation and individualism in order that firms may collaborate more effectively with their partners within the industry as well as from competing modes. Up till now, despite advances in micro-informatics, many small firms have access only to "off-the-peg" equipment when only a "made-to-measure" service is really effective. The increasing integration of micro- and macro-informatics will no doubt go some way towards remedying this situation".

In the long run one can expect that modifications to the supply and market structures in the European transport sector will occur. Presumably there will be only three different kinds of freight forwarders:

* The generalized group that comprises only a few international forwarders, as well as a few middle sized firms which have gained in size by cooperation (such as the company "Umzugs- und Transportsysteme" in West Germany).
* The group of specialists for specific tasks whose knowhow can be used directly for specific tasks. Their logistical concepts can be easily integrated into the major logistical system of generalized freight forwarders.
* The group of "suppliers" that supply the generalized firms with logistical sub-systems for their integrated logistical system. Sub-systems that become part of the overall logistical concepts comprise transport, storage and other services the generalized firms consider cheaper to buy. This group is dominated by haulers, the number of which will presumably increase since the freight forwarders will be more and more inclined to employ truckers. The deregulation of the European transport market will undoubtedly lead to a decline of road transport prices that will accelerate the structural changes in the haulage sector.

11.4.3 Result

The road transport sector will undoubtedly be able to meet the consignors' needs in the future. Due to the number of existing imponderables, however, the extent to which new structures will develop cannot be assessed. One of these imponderables is the exact impact deregulation will have in Europe. Since the introduction of logistical systems is a highly expensive matter, the extent to which the consignors' needs will be met depends on the commercial profitability of these systems. These unknowns, however, will have no influence on the principal advantages of road transport. Far better than any other mode of transport, road transport will be able to adapt to the growth and structural changes in European industries. Free of adverse regulations, the road transport sector will be able to grow faster than any other competitive mode.

REFERENCES

Bretzke (1987). "Entwicklungstendenzen von Ordnung und Organisation der Transport- und Logistikmärkte", Zeitschrift für Verkehrswissenschaft. Köln.

ECMT (1985). Foreseeable Cost Trends in Different Modes of Freight Transport. Report of the Seventieth Round Table of Transportation Economics". Paris.

IRU (1984). "Contribution du Transport Routier International à l'Expansion du Commerce Exterieur en Europe et dans les Pays Avoisinants Synthese de l'Etude." Genève.

Keimel, H. and Schmitt, A. (1981). "Szenarien der wirtschaftlichen Entwicklung der EG bis zum Jahr 2000." Köln: Deutsche Forschungs- und Versuchsanstalt für Luft- und Raumfahrt.

OECD (1979). "Facing the Future." Paris.

OECD (1986). "Technico-economic Analysis of the Role of Road Freight Transport." Paris.

Prognos (1987). "Ursachen, Ausmass und Auswirkungen unterschiedlicher Wettbewerbsbedingungen im europäischen Binnengüterverkehr. Marktsituation im internationalen Güterkraftverkehr und Bedeutung staatlicher Wettbewerbseinflüsse." Basel.

Chapter 12

THE INFLUENCE OF ADVANCED TECHNOLOGY ON FUTURE TRUCK DEVELOPMENT

Stig Ericsson
Saab-Scania, Sweden

12.1 INTRODUCTION

Today's technical development is forging ahead at a pace that would have seemed incredible only a few years ago. Many of us find it difficult to grasp and follow events in high technology - computers, micro-electronics, new materials, biotechnology and so on. How can this technical development be of any interest to something as mundane and unglamorous as trucks and buses?

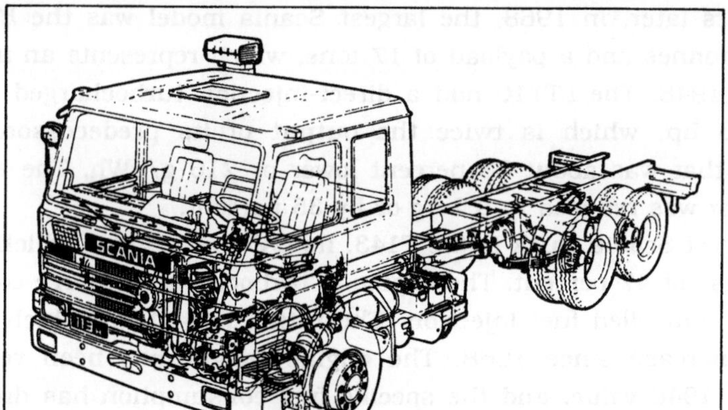

Figure 12.1

Truck engineering is a mature technology with a century of evolution behind it. So can we expect any revolutionary news? It all depends on our expectations on the degree of novelty. But developments in high technology also open new opportunities to truck designers, provided that we know what our customers need and what they demand of our products, and provided that we are sufficiently curious, interested and creative to put the new technology to use and make our seemingly simple products more advanced and better suited for the transport duties for which they are intended.

Even the already mature truck engineering is undergoing development day by day. But to assess such a development pace correctly, it is usually best to glance back a few years, and then try to extrapolate our findings into the future.

Let's therefore take a look at how trucks have been developed over a couple of 20-year steps, starting with the 1948 model from the time just after the Second World War, then considering the 1968 development stage, and finally the present-day trucks of 1988. What has happened during these past forty years, and what can we expect during the next ten or twenty years?

12.2 TRUCK DEVELOPMENT, 1948-1988

In 1948, the largest truck manufactured by Scania - the type LS20 - had a GVW of 16 tonnes, which is equivalent to a payload of about 10 tons. The truck was powered by a pre-combustion chamber diesel that developed 135 hp and weighed about 5 kg per horsepower. Its specific fuel consumption was 280 g/kWh.

Twenty years later, in 1968, the largest Scania model was the LT110, with a GVW of 25.8 tonnes and a payload of 17 tons, which represents an increase of 70 percent from 1948. The LT110 had a direct-injection turbocharged diesel with a rating of 275 hp, which is twice the output of its predecessor, and a fuel consumption that was about 23 percent lower or 216 g/kWh. The engine weight per horsepower was reduced to 3.2 kg or by 35 percent.

Today's largest Scania truck is the T143. Its GVW is 38 tons, which represents a further increase of 47 percent. The engine is turbocharged, charge-cooled and has electronically controlled fuel injection. The rating is 470 hp, which represents a 70 percent increase since 1968. The engine weight has been reduced to 50 percent of its 1948 value, and the specific fuel consumption has dropped to 191 g/kWh, or by 32 percent since 1948.

All of these figures might be difficult to grasp but the tremendous development of the diesel engined truck during the past 40 years will become evident when illustrated graphically as shown below.

This development would obviously never have materialized if other mature technologies had not been developed simultaneousy, such as the technology of building roads and bridges that could carry heavier vehicles. Developments that have taken place and are still continuing in these and other fields of mature technology are not all that far behind today's glamorous high technology, if you consider their importance to transportation around the world.

Against this background, let me attempt to extrapolate developments one or two decades ahead to demonstrate that high technology is one of the keys to the future. Let's start by taking a look at some of the main components of the truck and consider what developments we can expect.

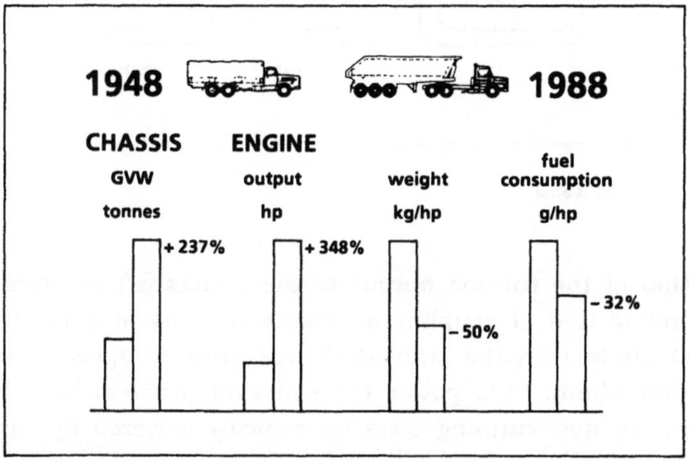

Figure 12.2

12.3 DIESEL ENGINE DEVELOPMENT AIMS

One such main component is the diesel engine which has now achieved an efficiency of about 44 percent. Most of the remaining energy is lost with the exhaust gases, and this particular loss amounts to about 35 percent. The heat lost to the cooling system accounts for a further 20 percent or so. It is these losses that we are trying to reduce, and some of the opportunities available are already undergoing practical tests.

In an attempt to recover some of the energy in the exhaust gases, we have installed an additional turbine in the exhaust pipe downstream of the turbocharger turbine. The power developed by this second turbine is supplied directly to the crankshaft across a special gearbox. This technique is known as turbo-compounding, and the improvement achieved so far corresponds to a reduction of about 5 percent in the fuel consumption, a gain of 10 percent in power of the same basic engine and very much improved torque characteristics at low speed.

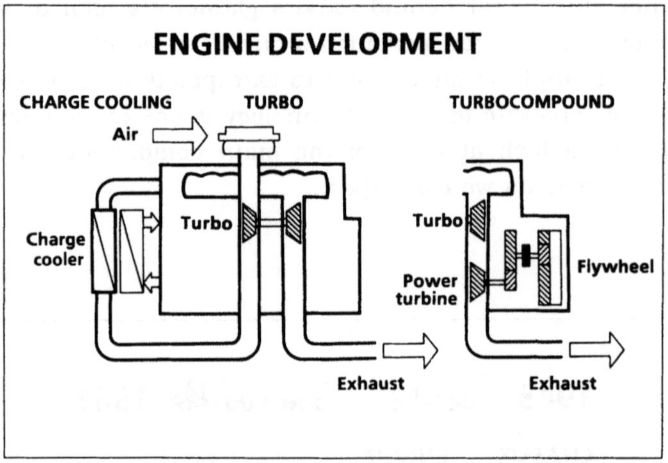

Figure 12.3

The transmission of the turbine output to the crankshaft involves a number of problems, including that of coupling a turbine running at a maximum speed of around 50,000 r/min with the crankshaft that runs at speeds varying between 1,000 and 2,000 r/min. This power transmission problem has, however, been solved, and we are now running tests on vehicles powered by turbo-compound engines.

Research and development in the field of materials is being pursued in parallel with work on the turbo-compound technique. In the case of engines, we are working on materials that have better heat resistance than conventional materials. Ceramics, for instance, would make it possible to insulate the engine combustion chambers, which would result in higher exhaust gas temperatures. This would enable more energy to be recovered efficiently in the compound turbine, and we could achieve, at least theoretically, an engine efficiency approaching 50 percent.

In recent years, microelectronics have made inroads into the engine, initially for controlling the fuel injection.

This area offers great development potential with regard to fuel consumption, shaping of the torque curve etc. Electronics will also have a beneficial effect on exhaust gas emissions, since a number of emission-related parameters can be controlled more accurately. It can even be said that electronic control is indispensable for achieving the limit values that will probably be specified by law from the mid-1990s.

Intensive work is continuing on the development of the combustion chamber shape, and on the performance and design of the injectors. All of these technical improvements would be very difficult to achieve without the currently available sophisticated technical design tools and laboratory equipment, largely controlled by computers and backed by advanced sensor and control technology.

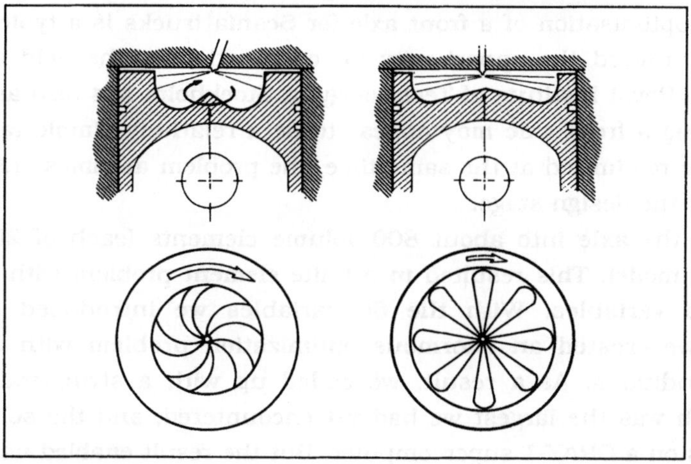

Figure 12.4

As I mentioned earlier, we can envisage efficiencies approaching 50 percent, but we must also bear in mind in our development work that every improvement must be of practical use to the vehicle owner. Maintenance is also an important element in this context, and efficient service must be available on all markets on which diesel engines of the new generation will be in practical operation.

12.4 INTEGRATING AIR TECHNOLOGY AND VEHICLE DEVELOPMENT

Certain parts of the increasingly advanced equipment and expertise available within the Saab-Scania Group have been coordinated with corresponding units within the Aircraft Division and the Saab Car Division. This form of cooperation within the Saab-Scania Group has proved to be beneficial in many respects.

In the first place, we at Scania have gained tremendous benefit from the vast expertise in the Aircraft Division in fields such as aerodynamics and strength of materials. In the latter case, this is particularly true in the area of advanced materials. But there are also other fields in which automotive engineers and aircraft engineers can contribute to each other's work. The fact is that aircraft engineers are at least one decade ahead of automotive engineers in fields such as materials technology, design engineering, computation, electronics and human technology. Automotive engineers have benefited greatly by this expertise available within our company. In return, we can contribute experience and knowhow in areas such as production engineering and production economics, where we probably face a more competitive climate than our aircraft colleagues.

The weight optimisation of a front axle for Scania trucks is a typical example of how we have utilized the opportunity for cooperation in the field of strength of materials. The Royal Institute of Technology in Stockholm has also assisted in this work. Designing a front axle may appear to be a relatively simple task, but if the weight is to be minimized at the same time, the problem assumes entirely different proportions at the design stage.

We divided the axle into about 800 volume elements (each of 20 nodes) in a finite element model. This resulted in a finite element problem with about 15,000 undetermined variables. With the 60 variables we introduced as the basic parameters, we created an enormous optimization problem with about 10,000 secondary conditions. As a result, we ended up with a structural optimization problem which was the largest we had yet encountered, and the solution took us 10 CPU hours on a CRAY-1 supercomputer. But the result enabled us to reduce the weight of the front axle by 15 kg. The saving is estimated to about SEK 3 million in reduced material costs.

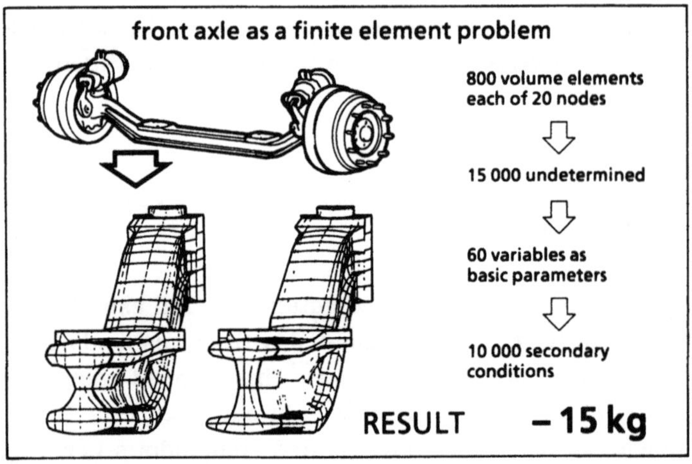

Figure 12.5

12.5 NEW MATERIALS

Developments in the field of materials are a topic of current interest to the trade press and other media. The prime-movers in this area are the aircraft and space industries which make strict demands on high performance and low weight, but need not necessarily take highly restrictive price considerations into account.

A great deal of basic knowhow on the strength and temperature tolerance of various materials and the related manufacturing techniques has now been accumulated. This applies, for instance, to carbon fibre materials which are currently being studied in the automotive industry. Numerous plastics and other non-metallic materials have traditionally been used in automotive engineering. But the areas now being studied concern more demanding components, for which the required low weight must be balanced against costs and other aspects related to operating economy. So it is hardly surprising that the publicity given to some inroads made by new materials in automotive engineering has soon faded, since the projects did not prove to be sufficiently down-to-earth.

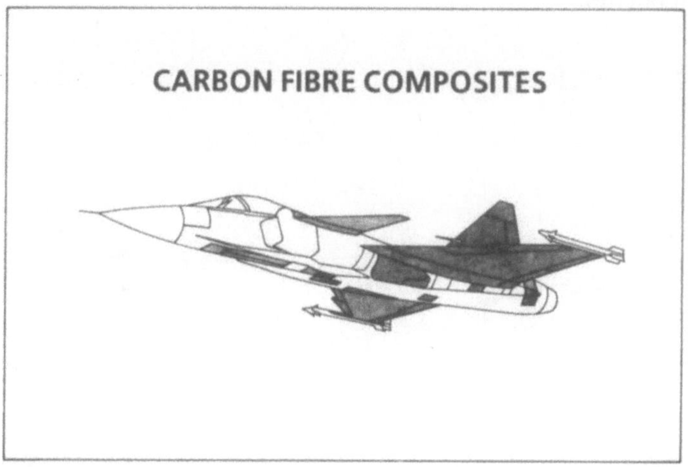

Figure 12.6

As an example, let me describe our development work on replacing steel by glass fibre composites for the leaf springs in the vehicle suspension system. An initial assessment indicated that the weight of a distribution truck could be reduced by around 200 kg, which would be an excellent sales argument. According to our specifications, a glass fibre composite spring should have suitable spring characteristics, adequate resistance to delamination, adequate lateral rigidity horizontally and high resistance to physical damage. In addition, the material and manufacturing costs must be such that the springs will be an attractive proposition to the truck owner.

But while the development of composite springs was in progress, better and lighter steel springs also emerged. In addition, the demand for air suspension accelerated. This sequence of events illustrates the common pattern of how advanced product development must often be aimed at moving targets.

Although we have access to databases which contain basic data on materials and also include field experience, many questions remain to be answered. We also have a great deal of knowledge of how epoxy and glass fibre composites behave after environmental exposure to high and low temperatures, moisture, erosion, solar radiation and so on. Expertise is also available in manufacturing methods, where increasingly sophisticated systems and computerisation are used in the production of various components.

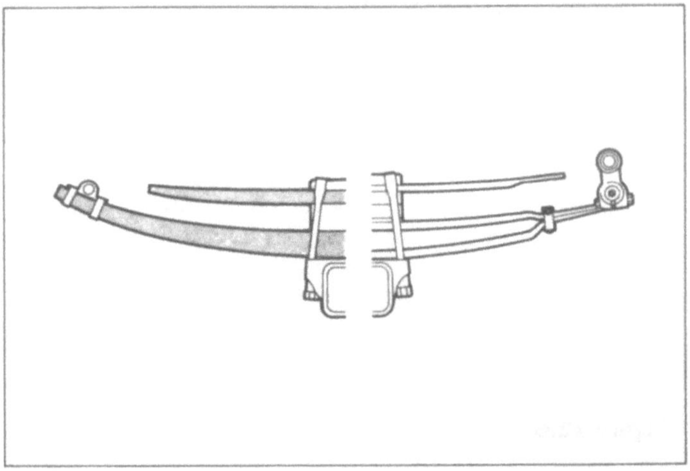

Figure 12.7

The desirable lightweight composite spring is thus attainable when considered in purely technical terms. But it remains to be seen whether or not it will become a commercial success.

12.6 BRAKE SYSTEM DEVELOPMENT

A development area in automotive engineering which is somewhat outside the interest of the press is the brake system. But this by no means indicates that we are passive on this front - quite the contrary.

The development of the payload capacity of a truck and the development of brakes can almost be said to be on a collision course. The demands placed on brake designers are growing enormously in pace with increasing engine power, improved roads, increased GVWs, decreased air drag and rolling resistance; the latter based on improved tyre technology which is striving towards smaller wheels with higher load-carrying capacities. The problem to solve is not made easier by further legal requirements that are beginning to prohibit the use of asbestos in the brake linings. High road safety and long service life are of course primary requirements.

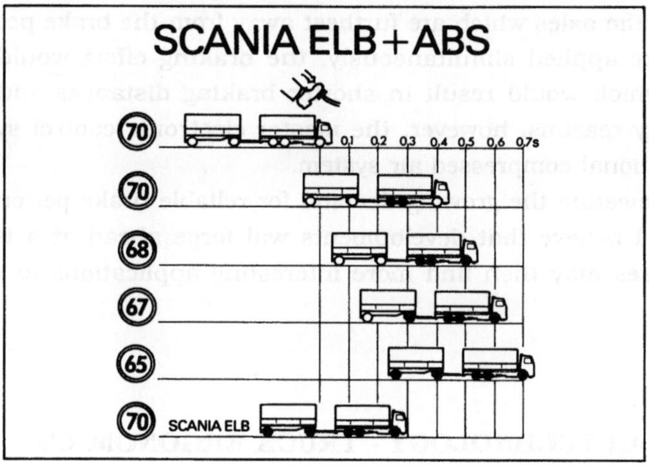

Figure 12.8

Disc brakes dominate today on cars and light trucks, but they are not equally self-evident on vehicles with axle loads in excess of 10 tonnes. Development work is in progress and we are testing disc brakes. Drum brakes are, however, still the most competitive alternative.

The integration of electronics into various functions of the brake system is a major development step. A few years ago anti-lock brakes were introduced. The benefit of this system is the significantly improved brake performance when the friction between the tyres and the road surface varies, such as on icy roads. Improved interaction between the brakes of different wheels also ensures more uniform brake wear.

For a number of years we have been working on a more advanced brake system. This consists of a combination of complete electronic control and pneumatic actuation. The system is code-named the Scania ELB system, and one of the objectives is to use electronics to ensure that the signal from the brake pedal is applied to all the wheel brakes simultaneously. The anti-lock, ABS, and anti-spin, ASR, systems will be integrated in the complete system.

One of the current difficulties which the ELB system is designed to eliminate is that braking of the axles which are furthest away from the brake pedal is delayed. If the brakes were applied simultaneously, the braking effort would be distributed more evenly, which would result in shorter braking distances and reduced brake wear. For safety reasons, however, the master electronic control system is backed up by a conventional compressed air system.

We are also meeting the growing demand for reliable brake performance by using retarders, and I believe that developments will forge ahead at a fast pace in this area. Disc brakes may then find more interesting applications to heavy trucks as well.

12.7 HUMAN TECHNOLOGY - TRUCK ERGONOMICS

Over the years, appreciable development has taken place in human technology. We usually refer to this as truck ergonomics, and the orientation applies to a wide range of factors affecting the driver's working conditions.

Ergonomics involve work on an interdisciplinary level. In this area, the expertise accumulated by the Aircraft Division is of benefit to us. Particularly important areas affecting the alertness and performance of the driver are the climatic conditions in the cab, noise level, level of physical stress caused by vibrations, and the ergonomic functions of the controls. The design of the driver's environment is directly related to cab development, which also involves strength and aerodynamic design considerations.

Cab development concerned purely with its strength is now carried out in advanced test rigs. We can simulate practically any road conditions, and we can also amplify these stresses to obtain results quickly. We can thus compress 5 - 10 years of practical operation into a few hours of computerized rig tests. In addition to this we are beginning to use advanced computer programs, which need supercomputer power, to calculate and dimension the cab structure for collision loads, etc.

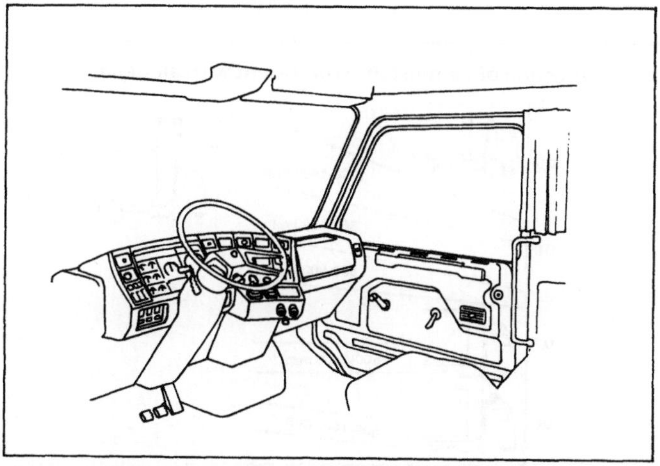

Figure 12.9

12.8 AERODYNAMIC DRAG

Aerodynamics experts from the Aircraft Division have assisted us in the aerodynamic design of the prevailing Scania cab. This turned out to be one of the most exhaustive studies ever undertaken on trucks. The study took more than 4 years and included numerous wind-tunnel tests on half-scale truck models.

One of the aims of this study was to acquire better knowledge of the influence of aerodynamic drag at different wind directions, yaw angles, in relation to the vehicle. In addition to resulting in a more efficient design of the cab itself, this knowledge has also led to separate air deflectors which interact with, for instance, a van body or a semi-trailer, with the aim of reducing the total aerodynamic drag. Briefly, we could say that the result of this study with regard to the reduction in fuel consumption corresponds to 10 years of evolutionary diesel engine development.

A great deal more could and must done in this area in the future. Work is continuing on other projects, so that we will be able to further reduce the aerodynamic drag, thus lowering the fuel consumption and the exhaust gas emissions.

The examples I have mentioned represent merely a modest selection of all the areas in which we are dependent on advanced technology and equipment for further development of vehicles for heavy transportation.

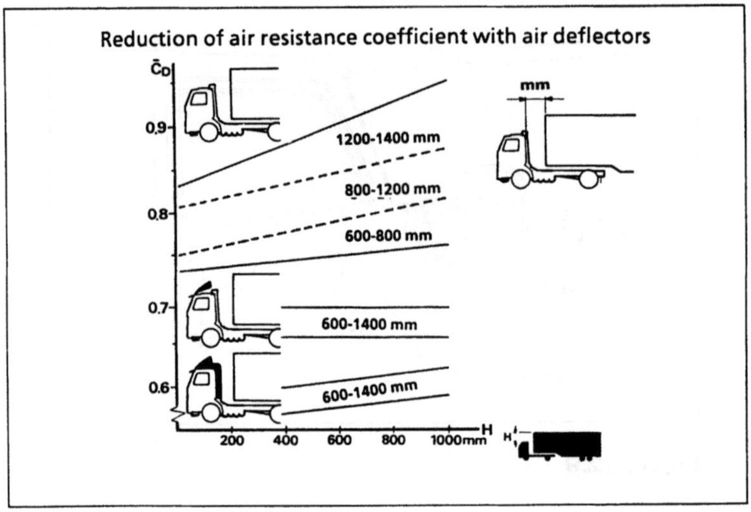

Figure 12.10

12.9 COOPERATION AND ENTHUSIASM

Finally, let me revert to the cooperation I have mentioned between the Saab Aircraft Division, the Scania Division and the Saab Car Division. This has quickly developed into more than merely project-oriented interchange of technological knowhow. The need for contacts and opportunities for getting to know one another has led to regular meetings between the development and production managers of the different divisions. The divisions act as hosts on a rotating basis, and an agreed and well-prepared subject is discussed during every meeting.

The subject areas which we discuss and which will be fruitful to future truck development to a greater or lesser extent include the development of steels and other metallic materials, such as magnesium, and the use of fibre composites and microelectronics. We also discuss basic matters such as CAD technology as a design aid, advanced computation techniques on supercomputers etc.

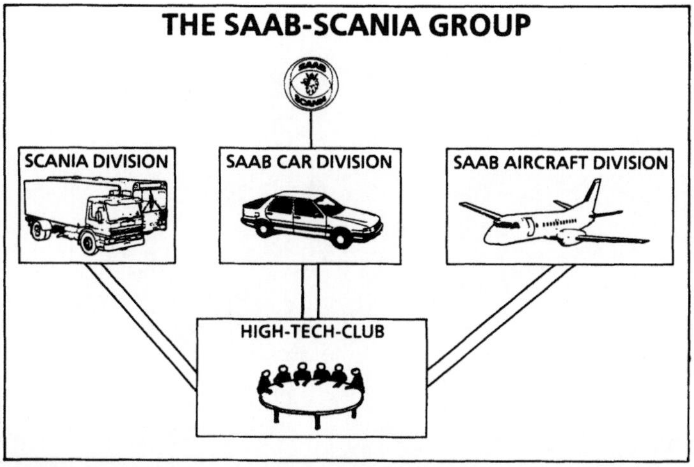

Figure 12.11

These advanced technology meetings - or high-tech meetings if you wish - have not only resulted in more efficient utilization of the knowhow and resources available within the Saab-Scania group. The personal contacts between the participants have created a sense of friendship and a team spirit which have turned the meetings into a form of club for positive and enthusiastic engineers, which I think is a prerequisite for their success.

Other companies with the same opportunities and problems as ours may possibly also follow this model for exchanging experience. In our opinion, it is one of many opportunities for the utilization of new technology to achieve even better results in preparing the truck for the next century.

THE FUTURE OF SURFACE TRANSPORT

(c) Passenger Transport and Commuting

Chapter 13

PASSENGER TRANSPORT TRENDS

Karl Krell
Highway Research Institute, Federal Republic of Germany

13.1 REFLECTIONS ON THE EFFECTS OF THE INCREASE IN THE APPLICATION OF INFORMATION AND COMMUNICATION TECHNOLOGY ON TRANSPORTATION

13.1.1 Passenger Transport

Information and communication technology is conquering increasingly larger markets with its dramatic growth rate. The driving force behind this development is presumably the improving cost-efficiency relationship, but also the play instinct of human beings which is stimulated by the interesting possibilities offered. Developments in the telecommunication sector already begin to irritate the planning experts who are concerned with the operation and improvement of passenger and goods transport.

Will telecommunications result in an increasing reluctance to travel? Will pupils and teachers in the foreseeable future no longer need to drive or go to school? Teleinstruction from home to home makes it possible, and written homework can be transmitted by telefax. Will teleconferences using video monitoring equipment, enabling participants to remain at their desks, become popular? In the service sector, it seems possible that office workers work at home and transmit results to headquarters, customers or colleagues by the appropriate telecommunication equipment. Even sales via video are claimed to become the way of doing business.

If such visions were put into practice, the passenger transportation industry would fall on hard times; but environmental activists opposing motor-driven vehicular traffic (motor cars in particular) would be happy.

Will the change towards staying at home become the new reality for people once used to going out to make a living?

This author believes that it would be wrong to promote such a development, and he also believes that it will not take place. By indirectly communicating via machines, people would lose touch as individuals, gradually fading into the state of a non-person in their human relations. Pupils need to be in direct contact with one another and with their teachers to grow up properly and become mature human beings. There is no way of catching the true mood of participants in a tele conference, not even with the best video equipment available. When the need to go out or be on the move terminates, our physical and psychic health will suffer. Travel and human contacts are instruments furthering success and well-being. Both should not degenerate into leisure time prerogatives.

The author's belief that people will not cease to travel because of the influence of telecommunication is borne out by the fact that a noticeable reduction in passenger traffic has not resulted since the invention of the telephone (see also Chapters 5 and 6). Long-distance calls certainly have made some travel unnecessary. But this mostly applies to the trips once regarded with disdain and which now are gladly cancelled. Those trips which are desirable can now be more conveniently discussed, agreed on and organized.

The better and more up-to-date information yielded by information and communication systems will indeed have effects on the quality of passenger transport (provided that the required data is collected properly and that the information needed at a particular time is quickly and automatically selected from the wealth of data stored). The following effects were pointed out by Cerwenka (1987): Better road and traffic information enables

* better utilization of vehicles (evasion of road interruptions and other incidences, reduction of delays when using different modes of transportation on a trip, knowledge about unusual hazards on the way)
* better use of road networks in the actual situation
* more mileage on the same infrastructure or a greater radius of action through reductions of time losses.

Even though telecommunication and better information have led to some unnecessary trips being cancelled, they will also generate new travel desires. In addition, the increase in leisure time in the industrialized world will result in increases in recreational travel, accomplished mostly by car.

To sum up it can be said that information and communication systems will not result in reductions in private car traffic, passenger transport or mileage. But it is expected that information and communication systems are suitable means to reduce the disadvantages of road traffic if they are used adequately and well designed.

13.1.2 Traffic Information and Traffic Management Systems for Road Traffic

13.1.2.1 *Detection systems and decision systems as a precondition for traffic i nformation and traffic management systems*

All information systems and all traffic management systems must consider that more and more elderly people will participate in the traffic of the future. Elderly people are slower in their actions. They need more time to adjust to different sight distances and sight/light conditions. Elderly people cause more accidents at intersections and often do not observe the priority of others.

We have to help them and to increase the quality of passenger transport by better information and driving aids. It goes without saying that aids of this nature should be so designed as to benefit all drivers. Assistance can be rendered only if critical conditions become known in time and the information can be relied upon. Increasingly automatic detection systems covering weather and traffic flow conditions and the resulting incidents have been installed for this purpose.

On general motorways in the Federal Republic of Germany, we are installing an automatic detection system by inductive loops at an average distance of 4 km. In regions where hazards are frequent, the distance between two loops is shorter. There are basically two possibilities of processing the data collected by these detection systems to put them to use:

a) the data on any critical situation can be transmitted to drivers on their route ahead and on (quasi) parallel routes, and it is left to the drivers to draw the required conclusions; or

b) the data are transmitted to a decision system (computer) where the situation is converted into recommendations or directions for action and are then transmitted to drivers.

Procedure b) has advantages but is associated with acceptance problems.

13.1.2.2 *Communication with drivers by means of variable message signs (roadside or overhead signs)*

At dangerous sections of the network, where fog and multiple-vehicle-accidents occur very often, we are planning to install variable roadside or overhead communication systems displaying internationally agreed mandatory signs. The messages of these sign systems will reach all drivers - also foreigners - directly. The enforcement of the observance of mandatory signs is the task of the police. The observance of recommendations or directions transmitted by in-car-communication systems cannot be controlled by the police. Expensive roadside or overhead sign systems cannot be installed everywhere. For that reason it has become customary worldwide to communicate messages to drivers directly, either over the car radio or via other additional in-car-communication systems. It is important in this case not to overload drivers.

13.1.2.3 *Requirements with respect to in-car-communication systems*

Detailed information on screen or other optical displays does no harm at home or in parked vehicles. But complicated input on route and information directly addressing drivers cannot be accepted for reasons of road safety, since they distract the driver's attention from the road and traffic. Krell (1986) lists the following requirements with respect to in-car information systems to meet the situation:

1. Connection of the communication system to the running system of the car, so that complicated input or detailed optical displays (likely to greatly distract drivers en route) are not possible when the car is in motion.
2. A signal indicating that the system is operating and properly functioning must be provided.
3 An ON/OFF switch must be provided to enable drivers to indicate their readiness to receive oral messages (possibility to switch the system off). If a driver signals readiness, the normal radio programme must be suppressed any time traffic information is given.
4. Traffic information to drivers should be primarily transmitted by acoustical means. A REPEAT button should be provided to enable drivers to let the last message be repeated. Additional optical displays should mainly function as a means of confirmation. They should present simple information only.
5. A change in information should be indicated by an internationally agreed sound. If drivers should continue on the route, oral route information should not be given.

Two alternative information sound signals could be of advantage. For example, sound signal 1 to indicate compatibility of in-car information with the signposting and sound signal 2 to indicate that the desired compatibility does not exist. In the latter case, some additional oral information is required to explain the situation.

6. Oral route information systems must be designed based on the principles of the technology of perception (volume control; announcement; comprehensible, brief, positively phrased sentences, etc.).
7. Optical display systems must be designed based on the principles of the technology and psychology of perception (placed at some central location, optimal angle of vision, brightness control, reflection-reducing cover, simple

arrow symbols, internationally unified, easily comprehensible signs or symbols, etc.).

En-route information should be restricted to a minimum but should not become extremely rare events (problem of vigilance). En-route information not relevant to road safety should not be transmitted where the full attention of drivers is claimed by other urgent matters.

The in-vehicle-route-guiding-system ALI-SCOUT, tested in Berlin (LISB) in 1988-89, already meets these requirements. The data exchange between the infrastructure and the vehicle operates via infrared beacons on the road and transmitter/receiver units behind the rearview mirror in the vehicle.

13.2 THE FUTURE OF PASSENGER TRANSPORT WITH SPECIAL REGARD TO COMMUTERS

As already mentioned earlier, the author does not expect a worldwide reduction in the number and length of trips/person/day to take place. Even though the number of trips by each citizen during a day will not increase considerably noticeable increases are forecast with regard to trip-length. Recreational travel is expected to rise in the industrialized countries. The number of journeys to and from work will not deacreae measurably as a result of the use of telecommunication systems, but as a result of a reduction in working days.

Which transport services will be needed in the future?

13.2.1 Rail Mass Transit or Private Cars?

At the present time,. the car as a mass phenomenon has come under criticism because of its impact on the environment (exhaust gas pollutants, traffic noise, severance effects of traffic streams, impact on pedestrian and bicycle traffic, visual intrusion), and the real and assumed danger of accidents associated with it. Car makers have hesitated to introduce the improvements which are technically feasible and would render cars ecologically less harmful. This has provoked public opposition to car traffic. In a state of dilemna, citizens are irrational in their behavior: they buy more and more cars which they want to use whenever and wherever they please, but object to others driving or leaving their cars in their neighborhood. They also object to new roads.

In this situation it is unclear whether suburban and urban traffic will be accomplished by car or public mass transport (PMT). If there remains a free choice of transportation mode, car traffic will be restricted only by congestion and parking possibilities. In this case the attainable proportion of PMT is not very high. Experts agree that this proportion can only be increased by restrictions on car use. The diagram in Figure 13.1 shows the attainable proportion of suburban and urban traffic using PMT means as a function of the traffic/environmental policy.

In our cities - especially residential areas - attempts are currently being made to restrict car use in certain areas (for certain local restrictions see Figure 13.1, second line). In particular, the unreasonable use of the car - by narrowing roadways, constructing traffic or speed retarders is being curtailed (Krell, 1974).

The expected growth in passenger transport (reference is made to this in Chapter 14) will primarily be felt on rural highways. The mileage in cities is expected to show a downward trend. The environmental activists call for the use of electrically driven rail PMT to eliminate these harmful effects of cars. Rail PMT is claimed to have the following advantages:

* In cases of high traffic demand, the space needed for the moving traffic is low. (With average demands, buses require less space than trains and if the occupancy is low, car traffic may even be the better solution.)
* Space needed for parking vehicles in the city centre is low. (But this requires more movements of empty trains to and from the train parking areas.)
* At present, the low requirements of primary energy outperform the car. (There may however be a chance of reducing the difference. In view of the occupancy of buses in the Federal Republic of Germany today, buses already require much less energy for the production of a passenger-kilometer than trains do.)
* Less air pollution due to the lower amount of harmful emissions and none at all along the tracks. That is why the placement of trains in tunnels does not require high additional costs for ventilation. Tunnels reduce the surface space required for traffic still further.

However, satisfying the claims of the environmental activists involves problems:

1) To meet the needs of commuters, mass transit requires a high concentration of workplaces or other trip targets, and also a high concentration of homes within walking distance of the stations.
2) Mass transit vehicles are in general fully occupied only twice a day, in the periods of demand peaks. Variations in the length of trains and schedules adjusted to the variations in demand levels pose problems. The empty trips to and from the depots or destination points cost money. Personnel expenses are generally the highest cost factor involved. The costs for the production of a passenger-kilometer of rail PMT in suburban and urban traffic is therefore higher than assumed-especially if the service is operated by public authorities.
3) PMT is not sufficiently attractive to people accustomed to their cars.

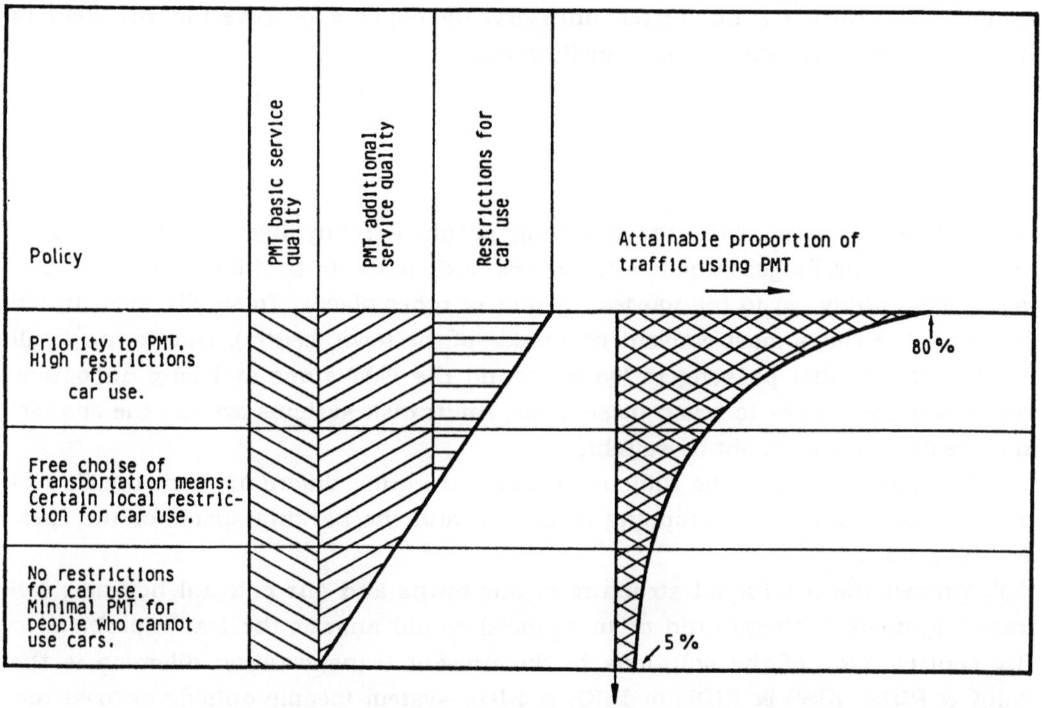

Figure 13.1 Attainable Proportion of Traffic Using Public Mass Transport (PMT) Means in Urban and Suburban Traffic (after Hamburg Consult 1988)

When the railways were the only means of high speed transportation, the concentration of building developments around stations referred to under 1) existed in most cases. The car has changed all that. In residential areas it has led to deconcentration and dispersal, dwindling residential densities, and homes far away from stations or stops. Workplaces also were decentralized. The car has made this possible. In short: under the influence of the car, urban development underwent changes no longer complemantary to the purposes of mass transit. Furthermore, people accustomed to their cars dislike walking or cycling to reach a station, although it would certainly benefit their health.

Under present conditions, rail PMT in suburban and urban traffic within the Federal Republic of Germany attains a load factor of between 25 percent and 30 percent. The production costs per passenger-kilometer are nearly equivalent to those of a medium-sized car. Since the fixed costs of a car generally reach about 60 percent, transit fares should not be higher than 40 percent of production costs if car owners are not to be financially disadvantaged - as they pay the fixed costs of their cars in any case. Another question is the responsibility for the subsidies.

Public authorities are no longer unreservedly capable of covering the deficits incurred by those running mass transit services.

13.2.2 Rail Mass Transit and Car Use

Commuters are faced with the following problem: many live in private homes (single-family dwellings) outside city centres and commute to the centre of town to work, shop, study, go to the movies, theater or other places. To satisfy these traffic demands, the car is most convenient outside of the town (centre). Due to the small number of potential passengers living around the PMT stops and long unpopular footways if one wishes to reach these stops, rail transit systems are not the answer. In some cases buses might be suitable.

In the centre of town the difficulties faced are quite different: there the flow of vehicles maintaining safe stopping distances, and the parking space for the cars, are the problem.

To protect the traditional structure of our towns and our cultural heritage, rail transit systems (underground or in tunnels) would answer the traffic problem in city centres. One of the solutions to the present transportation dilemma is the PARK & RIDE, KISS & RIDE or BIKE & RIDE system (people outside of town use their cars or bikes to get to a conveniently located transit stop and continue from there by PMT). However, this offer is generally only accepted if the use of cars in town is restricted (traffic retarders, parking restrictions, road pricing, prohibitively high parking fees, etc.). This fact is illustrated by daily traffic jams and wrongly parked cars causing problems for passengers and cyclists as well as visual intrusion.

The author does not believe that the special people-mover systems propagated in the past decade (low or high-capacity cabin-systems) will play a significant role in passenger transport of the future. The energy requirements and the costs of such cabins are very high owing to their weight and some unavoidable empty trips. The possibility of rapid transfers of cabins from one line to the next is rarely solved satisfactorily. Apart from that, the public objects to the visual intrusion of the required infrastructure.

13.2.3 Requirements with Respect to the Environment for Increased Car Use in Urban Areas

Being aware of the reluctant acceptance of mass transit, and knowing that financing transport would be easier if the people cherishing their freedom were to provide the cars (including maintenance and care) required, the justification of directly or indirectly enforcing the use of public mass transportation and restricting the use of cars in a democracy in the long term must be queried. Alternatives need to meet the requirements below:

A. The cherished traditional structures of our towns and our cultural heritage may not be injured noticeably in the long term.

B. Harmful car emissions must remain below critical limits (this requirement is all the more important in view of the limited absorption capacity of the atmosphere). Energy consumption should not be higher than it is today. Car noise pollution may not exceed mass transit noise levels.

C. Road safety (and pedestrian well-being) must improve compared with the situation today.

D. The impact on non-motorized traffic, and the visual intrusion caused by parked cars in towns, must be reduced compared with the present situation.

If the decentralization of places of work, shopping centres and cultural facilities continues the resulting traffic demand is not too high (up to about 20,000 persons/h), and if the environmental requirements are satisfied, a direct transport service by car might be thought of as a possible solution.

To satisfy requirement A, the capacity of the existing roadways needs to be improved considerably. To satisfy requirement B, vehicles with extremely low emission levels must be employed. Noise levels must be low (car body and tyres) and cars operated in an ecologically acceptable manner. The use of noise-reducing pavement surfaces (reducing noise levels by a few dB(A)), is currently under study in the Federal Republic of Germany. To satisfy requirement C, special provisions are necessary on arterial routes (arterial roads in towns with a high traffic concentration still cause the highest number of accidents). To satisfy requirement D, automatically moved cars parked outside of town, or multi-storey parking facilities accommodating high densities of parked cars at acceptable time delays, are needed.

13.2.4 Possible Roadway Capacity Improvements Without Roadway Widening

Once car emissions and noise pollution drops below critical limits, the following possibilities of roadway capacity improvement without roadway widening might be considered for commuter traffic:

* a demand-responsive reversible-lane system accommodating the flow of vehicles observing accepted safe relative stopping distances (relative safe headways)
* in-car motorist information and diversion systems
* reduction of the present headways of successive vehicles.

13.2.4.1 *Reversible lanes accomodating vehicles observing presently required headways*

Nowadays, urban networks are normally signal-controlled. With the political will required, signals are mostly coordinated so that lanes are used to maximum capacity while giving priority to PMT systems on the road. Lanes of arterial roads are normally devoted to a fixed direction. There is a need for changing lanes during peaks of in-bound traffic in the morning and of out-bound traffic in the evening. Lanes devoted to a fixed direction waste street capacity. Under the reversible-lane system, the central lanes could be designated for inbound traffic movements in the morning and for outbound traffic during the afternoon.

In three-lane roads this can be accomplished by changeable signs at the roadside. Overhead signs displaying the red St. Andrew's cross or a green arrow are a better solution. They are indispensible in the case of roads with five or more lanes. The change in direction naturally requires special provisions.

In addition there must be changeable message signs to inform motorists (prior to entering such roads) about actual lane directions. In certain cases intersections must be redesigned for this system.

Continuous road marking lines, instead of broken lines between lanes used for different directions or the placement of barriers or similar devices, would be desirable. But the positioning of such dividing elements on roadways can be a problem in countries with severe winters. Translucent elements inserted into the roadway and illuminated from below cause problems of skid resistance and also problems of visibility in bright sunlight or when the sun is in certain positons. The reversible-lane technique and its problems have been described in more detail by Behrendt and Krell (1988).

13.2.4.2 *Route guidance and warning systems by means of in-car communication*

Often the capacity potentials of parallel or quasi-parallel roads in a network are not fully utilized because of motorists lack of information about the lanes available and their utilization. That is why some routes are congested while other parallel routes are relatively free of traffic. The capacity potential of urban roads is further reduced by unnecessary and wasted trips of motorists searching for a destination or a parking space. Often drivers have difficulties finding the best way, even in unobstructed conditions. Network changes (changes in one-way streets, capacity reductions or obstructions caused by road works or accidents) still increase the number of search trips. Changeable message signs and changeable direction signs are of limited effectiveness if there are unexpected changes in the network situation. Large sign assemblies or overhead sign structures are objected to because of their visual intrusion into the townscape.

On a worldwide basis, considerable effort is being invested in the development of in-car communication systems to help motorists

* to reach their destination safely thus eliminating the unproductive additional traffic caused by search trips,
* to better utilize the roads available thus eliminating the development of any avoidable congestion,
* to avoid accidents because drivers experience no problem in finding their way and are warned in advance of road hazards (accidents, congestion).

The objective of assisting drivers in reaching their destination safely can be attained by an electronic road map in the car and automatic in-car information about the recommended route. Mapping alone, however, cannot take the current network situation into account.

To solve the latter, two approaches have been developed (Krell, 1987):

I. Electronic maps which can be updated by inputs into the main memory of a car via digital radiocommunication (e.g., via the radio-data system -- RDS) or via beacons; and

II. Transmission of all the current route and hazard data known from a control centre to cars via beacons (infrastructure controlled in-car communication).

The advantage of I. is that it could be implemented and introduced - on condition that policy makers agree to the in-car route finding algorithm - without additional investments in the road infrastructure. Infrastructure investments to some extent would be required by the digital radiocommunication system. With a mapping system of good quality, motorists could be assisted immediately and uniformly in finding their way. However, there is still the question whether the capacity of a digital radiocommunication system is sufficient to transmit the data required for updating information in due time.

With respect to II., there would be no updating problems. The effectiveness of this approach depends on the presence of beacons and an operational control centre. For that reason, motorists would not receive information everywhere during the initial implementation phase. This may be a handicap to its introduction.

Improved utilization of network capacity by means of in-car communication systems is often overestimated. As long as the demand for road space in cities exceeds the road space available, and people partially refrain from using their car because they fear the traffic jams and parking problems which are bound to occur, the possible improvements achieved in this way may be jeopardized by the fact that more people may wish to use their cars.

13.2.4.3 *Reduction of presently required headways*

A very different measure of achieving better roadway utilization is the reduction of headways between successive vehicles. At 50 km/h, a theoretical headway reduction of two following vehicles, measured from front bumper to front bumper, from 2 seconds (which is the safe relative stopping distance in this case) to half a second, and an increase in roadway capacity by a factor of 4 would be possible. Within the framwork of the PROMETHEUS study, such bumper-to-bumper driving or driving leeward of the car ahead is considered as being technically feasible. It is also feasible from the standpoint of safety with the use of modern electronics and direct communication between the in-car computers of successive vehicles. In line with the objectives of PROMETHEUS (attainment of unprecedented safety), the requirement of better safety under C would be satisfied simultaneously.

Deceleration lanes or turning lanes must be provided to eliminate traffic flow interruptions caused by vehicles wishing to leave the road at intersections. In this connection, it should be remembered that the capacity of urban roads depends entirely on the capacity of the intersections.

13.3 THE PARKING PROBLEM AND ITS SOLUTION

A special handicap to private car traffic - especially in commuter traffic - is the space required near the place of work to park the car during working hours (generally for periods from 6 - 10 hours). The office space required by an office worker generally amounts to about 20 m^2 (including corridors, WC, etc). A car used by this office worker would require about 25 m^2 (including approach ways). If all office employees required near-office space for their cars, the useful area of a multi-storey parking garage would have to be larger than that of the office building.

This problem gave rise to the idea of proposing the automatically - moved car, guided by some facilities in the roadway - which is automatically moved back to a parking facility outside of the centre once the driver and/or occupants have reached their destination. When again required, a signal will recall the car to the exact spot where its driver or occupants wish to retrieve it.

Automatic car movements are deemed to be technically feasible (they are also part of the PROMETHEUS study), but the associated safety problems (pedestrians losing their balance on the moving part of the roadway, etc) require very costly safety measures.

Apart from that, there is the realization that automatically - moved empty cars would cause more incidents in traffic flow. The main facts against this solution are the unproductive additional traffic generated by the moving of unoccupied cars, and the energy wasted by these empty trips.

This proposal does not solve the basic problem of land requirement; parking space outside of town will also be required if cars are not returned to their owner's garage.

If such additional parking space is required, one reasonable solution would be the design of good-looking multi-storey parking facilities, where only 12 m^2 per car and a low storage height is required. This is a task car makers should not leave to exotic inventors. A task of this nature might give a new impulse to the design of "park-efficient" cars.

Concidering the increasing crime rate in multi-storey parking facilities, provisions for safety would be needed. Attractive multi-storey parking garages are postulated (i) to remove parked cars from streets, sidewalks and bicycle lanes (reduction of hindrances and visual intrusion), and (ii) to create equivalent conditions both for mass transit and private cars.

In respect of the latter, public transport authorities generally assume that the reluctance to use mass transit vehicles is mainly caused by the walking distance to the next bus or tramway station.

13.4 CONCLUDING REMARKS

The solutions proposed cannot be realized without incurring costs, especially for governments. They may be partially financed them by the savings in subsidies earmarked for the mass transit service. However, zero-subsidies for public transportation will not be possible (see Figure 13.1), since it is part of the state's concerns to provide adequate public transport at a reasonable fare for those who cannot, may not or do not wish to drive a car.

REFERENCES

Behrendt, J., Krell, K. (1988). Fahrstreifensignalisierung (Signalling for lanes) in *Verkehrsleittechnik für den Straßenverkehr*, Band II, Springer Verlag Berlin, London, New York, Paris, Tokyo, p. 175-207.

Cerwenka, P. (1987). Auswirkung der Telekommunikation auf den Personen- und Güterverkehr (Effects of telecommunication on passenger and goods traffic), *Straße und Verkehr* 1987, p. 853-857.

Krell, K. (1964). Möglichkeiten zur Steigerung der Reisegeschwindigkeit von U- und S-Bahnen (Improvement to travel-speeds in Metropolitan Rail Systems), "*Der Stadtvekehr*" Vol 2, p. 46-50.

Krell, K. (1974). Gedanken zur Einschränkung der Kraftfahrzeugbenutzung (About the reducation of car use), *"Der Städtetag"*, 27/1974, H. 4, S. 217-224.

Krell, K. (1986). Kraftfahrzeuginterne Kommunikationssysteme (In-Car-Communication) Tagungsband der Forschungsgesellschaft für Straßen- und Verkehrswesen zum Straßenkongress 1986 in Würzburg, Kirschbaum Verlag Bonn, S. 106-110.

Krell, K. (1987). Bedeutung moderner Technologien für Ortung und Navigation im Straßenverkehr (Importance of new Technology for positioning and navigation in road transport) in *"Ortung und Navigation"* 2/1987, Deutsche Gesellschaft für Ortung und Navigation (DGON), p. 204-218.

Hamburg Consult (1988). Wirkungen von ÖPNV-Strategien auf den Modal-Split (Effects of traffic/environmental policy on the Modal Split) in *"Forschung Stadtverkehr"*, 1988, vol. 41, p. 65-67, Bundesminister für Verkehr, Postfach 200100, D-5300 Bonn.

Chapter 14

THE LONG TERM DEVELOPMENT OF PASSENGER TRAFFIC DEMAND: The German Example

Johannes Grevsmähl
Ministry of Transport, Federal Republic of Germany

14.1 THE GENERAL SITUATION

When we think about the future of our societies, and more specifically the future of transportation, we normally have the industrialized countries in our minds. These are characterized by a higher standard of living as measured by the average annual income per person. We also find high rates of motor vehicle ownership. Important differences exist with respect to population density. As can be seen from Table 14.1, there is a wide variation in population density between the countries mentioned, ranging from 19 inhabitants per square kilometre for Sweden to over 350 persons per square kilometre in the Netherlands.

The difference between industrialized and developing countries in income per head and vehicle ownership is tremendous (the table mentions Morocco as an example). This is also the case with respect to passenger traffic, where walking is still the prevailing mode in developing countries. I will not elaborate on this topic. However, it should be borne in mind that the world we usually think and speak of is only a part of mankind.

This chapter deals with the future of passenger transport in industrialized countries. Germany will be cited as an example simply because it is most familiar to the author. First there will be a short discussion of some factors that influence passenger traffic and their expected development. Then follows some information on the present situation in short-distance passenger travel, as well as in long-distance passenger travel, which can be used as a starting point for the discussion in Section 3 and 4 on possible future developments.

The material used is to a large extent taken from studies on transport demand for the Federal Ministry of Transport carried out by Prognos AG (Basel), Socialdata

(München), Deutsches Institut für Wirtschaftsforschung (Berlin) and Institut für Verkehrs- und Tourismusforschung (Heilbronn). Most of the material has not been published.

One factor explaining passenger traffic is, of course, population. The total population of the Federal Republic of Germany has been slowly decreasing since 1974. At the moment, this is a West German peculiarity since to my knowledge there exists no other country with a decreasing population in absolute terms. The decrease in number is accompanied by a distinct shift in the age structure. Younger age groups decrease in population, while the older age groups tend to increase in absolute as well as in relative terms. These tendencies are likely to continue into the next century and we may expect a significant reduction of the total pupulation with a continuous shift in favour of older people. Though the decreasing number is a German speciality, shifts in the age structure in favour of older people seem to be a general phenomenon at least in most Western European countries.

Table 14.1 Structural Charerateristics of Selected Countries [1)]

Country	Population Density Inhabitants per square km	GNP per capita DM[2)]	Private cars cars per 1000 Inh. 1976	1985
France	101	24 066	307	379
Fed-Rep of Germany	245	25 807	312	416
Japan	324	24 698	164	230
Netherlands	353	24 027	273	338
Sweden	19	23 181	351	377
United States	25	34 947	500	540
Morocco	51	719		24

1) Generally 1985 or the year most close to it.
2) Converted from national currency by using purchasing power parity rates.

Source: Federal Statistical Office, FRG

Another factor that can explain developments in passenger traffic is the regional distribution of population. In Germany, considerable differences exist in population density between conurbations and rural areas. For several years we have had a slow but continuous movement between these areas. Population as well as employment in conurbations tend to decrease relative to less densely populated areas and rural zones. Another change of land use pattern takes place within conurbations where both people and jobs tend to move from the centre to the margin. There is some discussion whether and how far that development is likely to continue, but a majority of experts think that it will. During the next few decades we may gradually achieve a more dispersed structure of dwellings, jobs, and other opportunities for passenger traffic. This again seems to be quite a common development for most industrialized countries.

In all industrialized countries we have high per capita incomes and a high standard of living. As can be seen from Table 14.1, the differences between various countries are small, even if one bears in mind the problems of converting monetary units from one currency to another.

Gross national product of the German economy has increased since 1970, with annual growth rates around 2 1/2 percent per year. Economic consultants working for the Federal Ministry of Transport expect the economy to continue to expand at a rate of somewhat more than 2 percent over the next two decades. Growth rates of about 2 or 2 1/2 percent seem to be small, but if the corresponding development of gross national product for a longer period is considered, it turns out that the increase in economic welfare is quite remarkable. In Germany, gross national product per head doubled in real terms between 1960 and 1985. According to the expectations of our consultants, it will again double between 1985 and 2010.

Motor vehicle ownership has reached a high level in all industrialized countries. It has been continuously increasing, the speed of increase tending to be higher in countries with less cars per head compared to the United States which still has the highest level of motorization. But even in the USA the stock of private cars has been increasing (see Table 14.1). The number of cars per 1000 inhabitants seems to vary between countries mainly because of the different length of time they have been accustomed to using motor cars: The United States longer than Western Europe, and Western Europe longer than Japan. There may be a small additional influence of per capita income. There is no recognizable effect of population density.

The high level of motor vehicle ownership (more than 400 cars per 1000 inhabitants in Germany) does not mean that all groups of the population have equal access to a private car. We know from a consumers' survey in 1983 that in Germany we had 8 million households without a car, i.e. one third of all households. We find households without a car mainly in the low income groups and the older categories of the population, whereas the younger and the richer to a high degree have a private car at their disposal.

Car ownership also varies between different types of region. This is reflected by car ownership statistics, and we also found it in a household survey on traffic behaviour that was conducted in 1982. From that survey we know how many people had a private car personally at their disposal, could make use of it within their household, or did not have it at all. In rural areas where the supply of public transport is poor and conditions for driving and parking are quite good, only 15 percent reported that they had no access to a private car. In conurbations, about 35 percent of households had no car available (see Table 14.2).

Table 14.2 Car Availability by Type of Region

Car is available	Conurbation	Rural Area
personally	34,2	39,1
within household	31,1	45,1
not at all	34,7	15,5

Source: Household Survey, 1982

The last general topic that should be mentioned is the environmental discussion in the Federal Republic of Germany. To many sections of the public, the questions of clean air, clean water, protection of forests, etc. are important issues. They have played quite an important role in policy making in general, and also in political decisions concerning the private car during the last few years. For example, we have had the introduction of lead-free petrol, and we have also had the first steps in the direction of stricter norms concerning exhaust gases.

14.2 PASSENGER TRAFFIC TODAY

According to estimates of the total passenger traffic in Germany, approximately 95 percent of all trips are short-distance trips with a maximum distance of 50 km. Roughly one quarter are trips from home to work, one tenth trips of pupils to school or of students. Another 10 percent is business traffic, 20 percent of the trips are for purposes of every day life such as shopping, and roughly 30 percent for leisure purposes. The following paragraphs will concentrate on trips generated by households. Business traffic will not be dealt with in detail. First I wish to present some information on short-distance traffic that illustrates some of the most important factors which determine mobility and modal choice. Then I shall present some information on long-distance traffic.

14.2.1 Short-distance travel

Most of our information on short-distance travel comes from household surveys which were conducted in 1976 and in 1982. In 1976, some 29 000 households were questioned, in 1982 approximately 21 000. Because of the design of the inquiry, we assembled information mainly on every day travel behaviour and we may use the results of the household surveys to say something about short-distance travel. In addition, I will report on some findings of a comparative study on travel behaviour (in the United States and Germany) that was jointly sponsored by the US Department of Transportation and by the Federal Ministry of Transport.

The most important factor influencing passenger travel today seems to be the private car. A person who has a car personally available makes three quarters of all trips by using that car. Another 17 percent is made by walking and 5 percent by cycling. Public transport turns out to be quite unimportant. By comparison, a person who does not have a car at his disposal makes 46 percent of all trips by walking, 14 percent by cycling, and for some 27 percent uses modes of public transport. According to the findings of our household survey, to have or not have a car makes a considerable difference to modal choice (see Table 14.3).

Table 14.3 Modal Choice and Car Availability - short distance travel

Mode of transport	Car is available: personally	within household	not at all
Walking	17,0	31,6	46,6
Cycling	5,3	20,9	13,9
Car	74,0	32,6	12,5
Bus	1,5	10,1	14,0
Other public transport	2,2	4,9	12,9

Source: Household Survey, 1982

Supply of public transport is, of course, much better in densely populated areas than in rural zones. One may think that this may stimulate the owner of a private car to use public modes of transport, at least in larger towns. As can be seen from Table 14.4, this is only marginally the case: the difference between the modal choice of a person who owns and can use a private car between conurbations and rural areas is quite small. There is a little more public transport in conurbations, but basically the modal choice behaviour remains unchanged.

Table 14.4 Modal Choice by Type of Region, for car personally available & short distance travel

Mode of transport	Conurbation	Rural Area
Walking	17,2	16,1
Cycling	4,2	5,5
Car	72,2	75,3
Bus	1,2	1,5
Other public transport	5,2	1,5

Source: Household Survey, 1982

The study comparing traffic behaviour between the United States and Germany had quite similar results. A person to whom a car is available makes about 90 percent of all trips with that car in the United States, and only in the very large towns of 500,000 inhabitants and more the share goes down to 86 percent. Public transport then substitutes the private car for a small number of trips. In Germany, roughly 75 percent of all trips of a person owning a car are made with that car, the variation between types of communities is not large and we observe a small favouritism to public transport only in the very large towns. Those in both countries who have no car available make many of their trips by non-motorized modes. In towns of between 100,000 and 500,000 inhabitants, 18 percent of all trips in the United States and 22 percent of trips in Germany are made with public modes. That share rises to 27 and 29 percent respectively in the large towns of 500,000 and more inhabitants.

In the United States, persons who have a car available make nearly twice as many trips per day than persons who have not. In the Federal Republic of Germany we have found a very similar relationship. The well known general assumption that the private motor car creates additional traffic seems to be confirmed by the empirical results of our household surveys and the material from the United States in the comparative study; since the empirical material includes walking and cycling there may really be additional traffic caused by the possession of a car.

The average trip length also seems to vary with car availability. From the comparative study some trip lengths can be calculated. In Germany, a person with a car has an average trip length of 11 km; in the United States of 13 km. Persons without a car have trip lengths of somewhat more than 7 or 8 km respectively. So in both countries the possession of a car favours longer trips. Beyond that, trips seem to be generally longer in the United States.

Other important differences exist between the groups of people having a car available and those who have not. For example, by definition all young people below the age at which a driving licence may be acquired do not have a car available personally. Their traffic behaviour is also different because of other reasons. The conclusion that the availability of a car leads to more trips and to longer trips is thus somewhat weaker than the conclusions about modal choice.

There are some other factors explaining travel behaviour. At least some of them should be mentioned in short. First we have the age structure of the population. Persons beyond the age of retirement, for example, make significantly less trips than do those in the active strata of the population. Employed persons make more and different trips than housewives. As a final example, I want to mention the position in the family life cycle. For example, single adults have a higher rate of mobility than married couples with small children.

14.2.2 Long-distance travel

As a source of information on long-distance travel I shall use a Consumer Survey conducted by the Federal Statistical Office in 1983. In addition to a wide range of other items, that survey also gathered some information on the journeys of private households. Because it was not a specialized transport survey, the term "journey" is not precisely defined from the view point of transportation. For example, there exists no definite minimum trip length. However, by and large the term "journey" may be interpreted as a long-distance trip. The survey gathered information from 50,000 households for the year 1983.

Regarding modal choice, the most obvious difference to short-distance traffic is that there are no non-motorized modes. When a car is available to the household, for long-distance travel the car is still the dominant mode but its share is somewhat lower. The railways, buses, and aeroplanes are also used by car-owning households. Households with no car normally use public modes of transport for long-distance travel. In Germany, the railways are still the most important public mode for these households (see Table 14.5).

Table 14.5 Modal Choice and Car Availability - Long distance travel

Mode of transport	Car available within household	Car not available within household
(own) Car	67,0	-
Railway	10,1	54,6
Aeroplane	5,9	5,6
Bus	11,2	22,5
Ship	0,6	0,6
Other	5,1	16,7

Source: Consumer Survey, 1983

In contrast to short-distance travel, for long-distance travel we have a strong correlation between the income situation of a household and various travel characteristics. Only 56 percent of households with an income below 1 200 DM per month make a journey at all, whereas 95 percent of households with an income above 10,000 DM do travel. From other sources it is also known that the number of trips per mobile household increases significantly with household income. So the figures mentioned in Table 14.6 may underestimate the connection between income and long-distance trip-making.

Households with a very low income often do not dispose of a private car. Accordingly, the share of private cars in long-distance travel increases considerably from the low to the high income groups. When a person with a low income travels at all, in the majority of cases he or she will use the railway. Railway usage decreases constantly with growing household income, whereas aeroplane use is increasing.

The Consumer Survey gives no precise information on the length of trips. From a broad distinction between different types of travel destinations we can see that with larger incomes there is also a marked increase in trip lengths. Journeys within Germany decrease from 75 percent in the lowest income group to less than 50 percent in the highest. Accordingly, there is an increase in international and longer journeys with rising income.

Table 14.6 Long Distance Travel Characteristics and Household Income (DM per month)

Characteristic	below 1200	1200-2000	2000-3000	3000-5000	5000-10000	over 10000
Mobile house-holds 1)	56,0	64,0	72,8	80,0	89,1	94,8
Mode of transport (% of journeys)						
(own) Car	16,7	25,4	45,4	61,6	65,9	64,9
Railway	41,5	36,2	25,7	14,1	10,5	10,7
Aeroplane	3,1	4,7	5,5	5,9	6,5	10,6
Bus	14,2	20,0	15,6	12,7	10,8	6,8
Ship	0,2	0,4	0,6	0,5	0,7	1,6
Other	24,3	13,3	7,3	5,2	5,6	5,5
Destination (% of journeys)						
within Germany	75,9	71,4	65,8	60,1	54,4	45,7
Europe and Mediterranean	23,3	27,6	32,9	38,8	44,1	51,1
Overseas	0,6	1,0	1,3	1,1	1,5	3,2
Number of journeys (in 1000)	4 247	10 361	17 312	34 287	25 944	2 662

1) Households with at least one journey in 1983 as a percentage of all households in the income group

Source: Consumer Survey, 1983

14.3 FUTURE DEVELOPMENT

It has been shown that today short-distance travel is clearly dominated by the use of the private car. The situation is, at least in some respects, astonishingly similar in the Federal Republic of Germany and in the United States. It may be assumed that this picture can also be extended to other industrialized countries. For long-distance travel, no such comparison has been made. Here one would be inclined to assume that the situation in other countries of the Western world may be similar to Germany.

How will this picture change in the future? I do not intend to give a quantitative forecast of the development of the number of trips, mode share, etc. Such a forecast for a single country would not be very useful in drawing general consIusions. My intention therefore is to discuss a number of factors that may influence future development, and then I shall draw some qualitative conclusions.

14.3.1 Factor: Car Availability

The first factor to be discussed is car availability. I want to give some information on driving licences by age group in the Federal Republic of Germany (see Table 14.7). Up to the age of 44, ninety percent of the male population possessed a driving licence in 1982 and, at least to the age of 34, eightyfive percent of the female population had a driving licence. From information about new driving licences one may conclude that (since the household survey) the young generation has acquired at least as many driving licences as we observed in 1982. There exist no reasons to assume that they shall acquire fewer driving licences in the future.

Concerning the older generation, we can assume that an acquired driving licence normally will not be lost. On this assumption, in the year 2000, ninety percent of all men up to the age of 65 and eightyfive percent of all women up to the age of 55, will have a driving licence. Until the year 2010, this percentage will creep up by another ten years on the age scale.

The discussion about driving licences demonstrates that we can assume that motorization will spread even further, mainly by the simple fact that still more people will become accustomed to using a private car. We will find the main groups of additional car users among the older men and among women above the age of 45. This trend towards more private cars will be supported by the expected rise of per capita income which, as was said earlier, may double again during the next 25 years.

This further increase of private car ownership will have significant effects, mainly on modal choice. For short-distance traffic we may expect a further decrease in walking (which diminished by about 6 percentage points between 1976 and 1982). Since the owner of a private car is most likely to use it whenever it seems advantageous to him, it is also likely that public transport will lose still more of its customers.

Apart from second and third cars in the richer households there will probably be additional (first) cars in households with low income. They will also use their cars for long-distance trips, thus reducing the market share of railways in the low income groups. Otherwise this increase in car ownership will not be very important for long-distance travel, because the one car available in the house-hold will normally be used for long-distance trips. The difference made by a second or third car will be small.

Table 14.7 Driving Licences by Age Group and Sex

Age group	Men	Women
18 - 20	63,9	54,4
21 - 24	87,7	83,6
25 - 34	91,2	84,1
35 - 44	90,0	75,6
45 - 54	85,3	51,6
55 - 59	78,4	32,9
60 - 64	72,7	28,3
65 - 69	65,8	16,3
70 - 74	61,1	11,4
75 - 79	43,0	6,4
80 +	27,3	5,3

Source: Household Survey, 1982

14.3.2 Factor: Economic Development

The basis of increasing economic welfare is increasing productivity engendered by technical and organizational progress. Increasing productivity can be converted into rising incomes or into less working hours for the average worker. During the past, we have usually had both. Average working hours per year have decreased in Germany by a rate of approximately one percent per annum. At the same time, we have had a considerable increase in average income.

In my introduction, I already noted the expected increase in per capita incomes. Our economic consultants expect that the trend of diminishing working hours will also continue, though with decreasing speed. They assume that the rate of annual diminution, will slow down to less than 0.5 percent after the year 2000.

Diminishing working hours will probably lead to a further spreading of the six weeks' leave which is already quite common in Germany. In addition, we may get the four-day working week around 2000 or a little later.

As has already been mentioned, rising per capita income will support the spread of the private car. In long-distance travel it will cause the number of trips per household and per person to increase, and these trips will tend to be towards remoter destinations than today.

The gradual introduction of a four-day week will, of course, lead to less home to work trips. What this will mean for transport depends on how the topical question of flexibility will be solved. If the individual four-day schedule could be spread over five or six days of the calendar week, this could considerably ease traffic conditions in peak hours.

Less working hours also mean more time for leisure activities. These will generate additional traffic. The trend towards more long-distance trips, caused by growing income, should be supported by the increasing availability of leisure time.

14.3.3 Factor: Land Use Pattern

The location of dwellings, workplaces, shops, and other facilities is an important factor to explain modal choice and traffic distances. In countries with a historically dense town structure, we have experienced during the last decades a continuous spreading over space; this process was stimulated and rendered possible by the development of modern means of transport, especially that of road transport. The private car made it possible to reach a workplace or a shopping place relatively far away in a relatively short travel time. So people could and did move away from the centres. The process was accelerated by bad driving conditions in the city centres themselves. Concomitant changes in land use pattern favoured the use of private cars, since only the private car could provide sufficient and quick transportation in the dispersed structure that gradually emerged. Public transport was unable to adapt. Walking became more and more difficult because of the growing distances.

The process that changes land use patterns in a direction which is favourable for the private car, and a private car that creates traffic conditions which are favourable for that direction of development, is a self-reinforcing one which seems likely to continue. Because of the long useful life of buildings and infrastructure, this process will be slow. But even slow developments may produce quite remarkable results if the time span is long enough.

A more dispersed settlement structure of dwellings and workplaces favours the private car as a mode of transport. In short-distance traffic, it will be more and more difficult for traditional forms of public transport to offer competitive services to traffic users at a reasonable price. New forms of flexible public transport are being studied but, to my mind, there has been no real breakthrough. Consequently, the share of public transport will probably continue to decline.

Long-distance travelmay not be affected idirectly. However, it must be remembered that public modes of long-distance travel only provide a service between stations or between airports. When land use patterns become still more dispersed, this may make access and egress more difficult. The private car usually provides a door to door service in long-distance traffic. The described development may give it an additional advantage over the public modes in long-distance traffic, especially for medium distances up to 200 or 300 km.

14.3.4 Factor: Quality of Transport Services

By the end of this century, the Federal Republic of Germany will have in operation some important new railway tracks, and several other tracks thatwill have been upgraded. In combination with new rolling stock, this will make it possible for the German railways to improve their long-distance passenger services considerably. In consequence the railways will attract additional traffic away from their competitors. Experiences with improved rail infrastructure in other countries, for example in France and Japan, support such an expectation.

Another area where changes in the quality of transport service may influence future developments is in short-distance public transport. Most public transport organizations have been making losses. The states or the local communities that own these enterprises are prepared to pay for those losses only to a limited extent. Public transport organizations will be compelled to rationalize in order to save costs. Rationalization will also mean they will close down lines where there are few customers. Downgrading this service will in turn mean that there is reduced demand which may again lead to additional supply reductions. In my opinion, the established negative cycle of less demand and less supply will most probably continue.

14.3.5 Factor: Population

As has been mentioned, Germany is experiencing a decrease in its population. It started with a significant reduction of birth rates over a ten-year period between 1965 and 1975 School children were the first group where the reduction of birth rates was felt. School children represent an important group of customers for public transport, so their reduction has intensified the problems of public transport; especially in rural areas where public transport has fewer customers anyhow. The reduction of trips from home to school because of decreasing numbers of children is likely to continue until 1990 or 1995.

Another effect of population development is a consequence of trip generation. Decreasing birth rates mean that more couples have no children, and also more couples have only one child instead of two or three children. We know from comparative studies on mobility that households with several (small) children tend to make less trips than young adults or married couples without a child. Because of the change in the number of children per family, we may assume that mobility in Germany during the last decade has risen more quickly than it would have with constant birth rates.

In the long run, the shift in favour of the older age groups will be important. By 1984, persons older than 60 represented twenty percent of the population. According to new population forecasts, their share will probably increase by the year 2000 to twentyfour percent, and by the year 2010 to twentysix percent. These old people will probably be more active than the corresponding age group today, and they will in many cases own a private car. It is likely that they will also create more traffic than old people today. They will not generate as many trips as employed persons in their thirties or forties, but they will be much more flexible in their traffic behaviour than during working age. The shift in population structure may thus slow down the expansion of traffic and at the same time ease peak hour problems.

The expected reduction in absolute numbers of population may ease the pressure on infrastructure in general. On the one hand, we can expect car ownership rates to increase. On the other hand, the number of people who will own these cars may decrease. It is possible that both tendencies will counterbalance during the next century.

14.3.6 Preliminary Conclusion

Let me now integrate all individual factors into a preliminary scenario. Car traffic will continue to grow because of increasing motorization, higher incomes, and changes in land use pattern that favour car traffic. In the nearer future, the growth we may expect will lead to an additional burden on infrastructure; which might be eased a little by a more proportionate distribution of trips over time if changes to more flexible working times are possible. In the more distant future, however, we must expect this expansion to come to a standstill. First, that part of growth which is due to further diffusion of the car to other groups of the population will come to an end within the next ten or twenty years. Because of the change in population structure we may expect an additional retardation. In Germany we will also have the effect of a decreasing number of people.

Short-distance public transport will lose still more customers. Its quality of service will deteriorate. For most people (the car users) this will pose no problem. But with full motorization there will remain certain groups that may not or cannot use a private car. They will find it more and more difficult to make the trips they wish to make. Our societies must be willing ensure that these groups, can enjoy a certain minimum level of mobility.

Air transport will benefit from rising incomes and increasing leisure time, and should continue its steady growth. Higher speed railways will attract additional customers for long-distance trips. On the other hand, the railways may lose some customers who will make these journeys with the car they will have newly acquired. On balance the railways may maintain their long-distance passenger traffic or even gain a small additional share of the market.

14.4 SOME POSSIBLE LIMITATIONS

About five years ago, the prevalent opinion among the general public was that oil shortages and rising oil prices could present a serious difficulty to any further development of car traffic. Today this is largely forgotten, illustrating just how quickly expectations about long-term developments may change. The stock of oil in the ground is a limited resource. Apparent affluence should not mislead us to assume that oil products will never be scarce and expensive again. Of course they will. When this happen is uncertain. The prevailing expert opinion of today suggests that it may be in the year 2000 or even later.

Because replacement by other sources of energy takes place earlier in other sectors of oil consumption, such as example heating, it is likely that scarce oil resources will be reserved for transportation (and some other sectors, e.g. chemistry) - thus prolonging the lifetime of oil use. We also know that alternative sources of energy for motor vehicles do exist, though probably at a much higher price than today's petrol. Energy will not be a serious obstacle to further car usage even if oil resources should be exhausted.

In Germany there is considerable concern about the environment and the negative effects that human activities have on it. Three years ago, the dying forests were the dominant topic. Then the focus shifted to the risk of radioactivity caused by defective power plants, the pollution of rivers by the chemical industry and the pollution of ground water by agricultural activities. Though these issues may be forgotten quickly, in my opinion there will be new issues every year. The continual emergence of new environmental problems will create a constant pressure on political institutions to introduce measures protecting the environment. We must expect that there will also be additional measures concerning car and air traffic.

These measures will not prohibit driving, but may put certain limitations on it (speed limits, restriction of driving in certain areas). They will compel producers to minimize the negative effects of driving and flying on the environment.

Another limitation to the development of car and air traffic may be the limited capacity of infrastructure. For some years in Germany, it has often been very difficult to realize new major infrastructure projects because of the resistance of local interests. In several cases, a new runway or a new motorway proposal had to be resolved by the courts in a legal proceeding lasting for several years. This resistance to new infrastructure investment is prevalent in other parts of Europe. It must be interpreted as an indication of a widespread objection to certain types of transport infrastructure. This is further reflected in the budgets of public investors, where we have witnessed a slow but continuous reduction in funds dedicated to road building in real terms.

Today, air and road traffic are running short of infrastructure capacity. Because of this, they suffer losses of quality. It is possible that the situation will deteriorate further with growing traffic demand, because infrastructure capacity will not increase as quickly as demand dictates.

14.5 A FINAL CONCLUSION

Now let me come to a final conslusion. Despite the above constraints, we may expect a growing transport demand for road and air travel. Short-distance public passenger transport is likely to decline, but long-distance railway traffic may achieve a slight increase due to improved service quality.

There are certain risks to future expansion which concern mainly car and air traffic. They will not prohibit further development, but suggest that the future of car traffic and air traffic should be planned with great care. In future, both modes of transport must be more energy-efficient, less dangerous for the environment, and make optimum use of all infrastructure.

PART IV

PREPARING FOR THE FUTURE

Chapter 15

TRANSPORT EDUCATION AND TRAINING: PREPARING FOR THE FUTURE

Joseph van Stappen
Conseiller général de l'Institut du Transport Routier, Brussels

15.1 INTRODUCTION

Various speakers before me in this conference on "Transportation for the Future" have naturally dealt with the future evolution of transport and their technological and economic visions and expectations in this connection, with particular reference to the transportation of goods, and primarily to road transport.

I am not going to repeat their observations, but will rather examine how transporters and all those affected in the transportation and related sectors can be prepared as well as possible for the new situation. This will be done by suitable education and training, based on research and on surveys.

In references to the forthcoming technologies of the future it has been said that as we are the ones who are to use them, therefore we must know how to use technology, how to handle knowledge. If we do not prepare ourselves to become 'knowledge handlers', we will run into one of the bottlenecks of the future which has been referred to during the conference.

15.2 TRANSPORT IN THE FUTURE

By way of introduction, I shall just sketch the major perspectives which can be discerned in the world of transportation.

1. In 1992 we will face a radical change, particularly in Europe. When the frontiers of the transportation market will be shifted by the construction of <u>a single internal EEC market.</u> This liberalization of the market is based on the dismantling of all sorts of quantitative limitations and on measures of

deregulation. This creates a broader market with new and greater opportunities for the transportation business.

In order not to disturb the market pattern, and in order not to let competition between transporters and between means of transport get out of hand, the progressive liberalization will be accompanied by a more or less appropriate harmonization in the fiscal, social and technical sectors. This, at any rate, is what the European authorities are striving for, but in view of the inevitable imperfections of any harmonization, it is still all vital for transporters to detect their competitive advantages and to exploit them. It calls for competent and adroit management.

2. The gradual evolution of economic events is at present going through a simultaneous and still more radical change. We stand before the beginning of a new era. After the pre-industrial era, which was characterized by mechanical methods of work, came the industrial era in which energy and its applications set the pace. Now we have landed in the post-industrial society, the era of information technology. We have reached a breaking point in the graduality of economic evolution, even in the transport sector, and this calls for an appropriate strategy. The people who are going to formulate and apply this strategy need to be trained for this task if we are to attain the maximum efficiency from the networks.

3. Transportation has become more than just moving goods from one place to another. The logistic concept determines the pattern of transport and integrates the flow of goods from the producer to the final consumer. It also determines the physical displacement and the informative control of the stay and the state of the goods at each intermediate moment. This demands a new strategy for transporters, with marketing insights that make them, as it were, into co-producers and co-sellers of the product.

   Thus training, or refresher training or retraining will be necessary for many people. As the logistic concept requires that more will be transported in smaller quantities, in line with the just-in-time principle, and this approach fits precisely into the line of flexible road transport, such training will be particularly important for road transporters..

4. Thus the organization of transport will become far more extensive and more comprehensive. The whole transport chain must be controlled with shippers and receivers in a continuous and integrated manner, with the aid of information systems and telematics. These tasks will be taken over by a new generation of transporters who will be organizers of transport rather than those who physically forward the goods. The process of adaptation is under way, but there is an enormous need for skilled people to carry it out.

5. Also, increasingly more attention is being paid to external factors. And rightly so, especially in respect to traffic safety. In the European Community, traffic

accidents kill 55.000 persons every year, and another one and a half million persons suffer from traffic injuries, including 150.000 who become permanently handicapped. The financial cost is about 3 billion ECU per year. Road transport partly bears the blame for this, though not the greatest blame by a long way.

None the less, there are some who point the finger at road transport, particularly where the transport of dangerous substances is concerned. Here, transport firms should, on the one hand, make an effort to promote safety measures and, on the other hand, take better care of this image. For both these efforts, training is needed at all levels.

6. Care for the environment is also receiving increasing attention in transport. Road transport is partly blamed in Western Europe for the increasing backlogs of traffic and traffic jams, although road transport is also a victim of these. The infrastructure needs of traffic and transport are being contested and rejected in some circles. Road transporters are being blamed for noise pollution and other assaults on the environment. Therefore the transporters, in consultation with all the other parties affected, and in the first place with the government, will have to work out compromises to reconcile the economy with ecology as far as possible. To do this, transport firms need training as skilled interlocutors, so that they can improve their image in this respect by appropriate management.

15.3 PREPARING FOR THE FUTURE

So it is established that in the near future transport will go through radical changes and that many of these changes have already begun. These changes will persist in all directions, and nobody is in a position to resist them, even if they wanted to. All that can be done is to follow this evolution to make sure that it progresses in the most efficient manner. In other words: the future must be prepared for in good time.

In the first place, this means that we must take care to see that there are people who are ready and able to foster the transport of the future. People at every level, in all branches, from top to bottom: Researchers, planners, and those who decide on the necessary infrastructure, taking adequate account of external factors and economic needs; Designers and builders of the new means of transport: vehicles which will be increasingly equipped with more "artificial intelligence" and forms of telecommunication; Transport contractors who act as organizers and managers of the logistic chain and design an appropriate strategy; Employees and workers who know how to handle the technological devices; A new generation of drivers of road vehicles who are also pawns in the strategic policy and the marketing of the transport firm. In a word, the new transport of the future calls for new transporters and people to reframe the business scene.

The answer to this challenge can come only from education and training. To the extent that today's education and training programmes are not adequate to meet the changed demand, we must now intervene so as to be able to supply suitable people. Education must adjust in time to satisfy the demand of tomorrow's business community, it must be linked to the labour market, in both quantitative and qualitative terms. "It is important to remember that the human resource is the most critical component of an organization's ability to accomplish its goals. Only skilled and motivated transportation professionals can lead the industry into the 21st century." (Meyer, 1985.)

Significient proportion of today's executive personnel is untrained or incompletely trained for the new tasks. So there is a serious need for refresher training, which is partly justified for social reasons.

More important, however, is the distant future and the preparation for it. Business needs a new type of worker to be trained, a "homo novus", or, as Professor Salomon (Northwestern University) so graphically represented him in a drawing, the "homo cabellus", a man whose brain is the terminal of the "cables" of our technological world with its "information systems". This third factor of production will control the business of the future.

A conference such as this which aims at examining and depicting what "transportation for the future" is going to look like, must address the question of how this future is to be prepared. Europe must make good its place in the world of tomorrow, and it must develop the necessary talents to do this, as it can no longer be regarded as an expansive region in the world. For, as Pasteur said, "le hasard ne favorise que les esprits bien préparés" (fortune only favours well prepared minds).

15.4 THE DEMAND SIDE

For a better analysis, we can look at "human resources" as a market problem with a demand side and a supply side. In such a theoretical approach to the problem, the important thing is to examine whether supply is adequately attuned to demand. In other words, whether education and other forms of training are delivering the suitable workers wanted by business and government - in this case in the transportation sector - and if this will also be the case in the future, in both quantitative and qualitative terms. Finally, if it appears that education and training are not meeting the demand from business, we should examine how supply can be better attuned to demand.

First of all, good statistical information is necessary, not only about present needs, but also about the needs of tomorrow. In forecasts over a period of at least 6 to 10 years for various European countries, it is not only quantitative information that should be provided but above all qualitative information. The transportation business will have to determine what training requirements should be set for "professionals" of transportation it wants to recruit. It must alo decide what refresher training possibilities it wants for personnel who are already employed and who have to adjust to the rapidly changing circumstances of business. For this purpose, market analyses and surveys are necessary, but there are still very few being carried out. Perhaps this conference can be an "eye-opener" and a spurt in this direction.

A few days ago, I happened to learn that in Germany and Holland a survey has been conducted in order to find out what the needs and requirements of the road transport sector are in the field of refresher training.

In Germany the B.Z.G.-Unternehmensberatung stipulated: "Planmässige Fortbildung in den mittelständischen Unternehmen des Verkehrsgewerbes zur Qualifizierung der Leistungen und zur Steigerung der Wettbewerbsfähigkeit" = "Additional (continued) training, in accordance to a preset plan, for small and middle size transport companies in order to reach improved quality in their operations and in their competition fitness".

In this German study it is stated that over and above the 3 traditional bases for professional qualification and competence which are:

- Education in the schools
- Professional training
- Experience gained on the job, --

a "continued training" is now indispensable as a fourth base.

This continued training must be planned in accord with preset objectives, at the short term, middle term and long term.

Therefore, a catalogue of required knowledge and professional skills is to be inventorised and compared with the present situation.

The German survey of a number of road haulage companies revealed the various needs for additional training, at several levels.

For the head of the road transport companies the needs lay primarily with:

- rhetoric capacity, that is "oral communication"
- management of "time"
- management of "people within the company", the staff members (Um gang mit Mitarbeitern): creativity management.

The Dutch survey was conducted by the NEA-Institute, the Dutch Institute for research, education and consultancy in the field of traffic and transport.

It also surveyed a large number of road transport companies and made a distinction between the shortages in knowledge in various fields and, on the other hand, the need for courses to improve skills and for courses to improve professional knowledge.

Shortage of knowledge are particularly sharply felt in the field of automatisation (with 55% of the companies) and in the field of logistics (46%) and marketing (43%).

88% of the companies said that additional training is the best way to overcome the shortage of knowledge, rather than recruiting additional personnel.

Where transportation is concerned, on the demand side we have to distinguish between a number of sub-markets. I shall in any case limit myself to the transportation of goods by road in Europe, the theme of today's session.

1. First of all there is the demand from the transportation business itself. In all European countries of Europe, the road transport sector consists almost exclusively of SMCs - small and medium companies. In some countries, the overwhelming majority are even small one-man businesses; in some other countries there are a few larger or mixed transport firms. Everywhere we see a trend towards economies of scale: in the number of vehicles being operated, in the load capacity of the trucks and of the vehicle fleet, in the volume and range of activities. At the same time, this development is accompanied by a certain specialization of services and the segmentation of the market, in order to be able to discover and exploit competitive advantages more effectively. The new challenges are just ahead of us and will determine the future: I have summed them up in six introductory points, following on the vision of the future given by the preceding speakers, namely the broader, open Euromarket, the application of information systems and telematics, the logistic concept with new organizational forms for transportation, and finally the external factors of environment and traffic safety.

   a) In the past little scholastic pre-training was expected or demanded of road hauliers. This has changed sharply in the last 20 to 30 years, and still more knowledge will be required in the future. In contrast to former times, more and more people with higher education are being employed in the road transport sector, and we see the sons and daughters of self-made road hauliers taking over the family business and running it with a university degree under their belt, together with educated collaborators in various departments.

      The profile of the head of a road transportation business wishing to ensure his survival is thus distinctly modified, and naturally the training must be correspondingly modified.

      Some countries made significant progress after 1974, when the Council of the European Community approved a Directive (74/561) on admission to the occupation of road haulage operator in national and international transport operations. The knowledge that was demanded from the future operators related to legal subjects (commitments, transport agreements, trading companies, social and labour regulations, taxation, etc.), to the commercial and financial management of a transport firm, to transport regulations (both national and international), to the technical standards for vehicles and for their maintenance, to road safety, customs practices, etc. These qualifications must be demonstrated in an examination, and in Belgium this examination has both written and oral parts.

A proposal for modifying this Directive is now before the EEC Council for approval. Among the proposals are that the qualification requirements should be extended and made more difficult in order to raise the standard of road transport in all EEC Member States again. At the same time, it is recognized that the intention is also to adapt what is taught to the needs of the future, by the addition of subjects such as logistics and management techniques, particularly in the spheres of strategic planning and marketing. Knowledge required in the field of traffic safety and environmental regulations will also be increased, partly with specific reference to the transportation of dangerous substances.

This already gives you some idea of how the professional profile for the future head of a road transport firm will appear. The qualifications required will vay in breadth and depth according to the size of the firm.

For transport firms aiming to expand on the new European market, the management requirements will be substantially higher than the basic conditions for access to the profession would indicate. Apart from the mastery of integrated logistic techniques, making use of information technology, character aptitudes and attitudes will play an increasingly important role.

Thus the qualifications demanded for the heads of road transportation firms are determined and described.

What we need to do is to translate the content of the profile into education and training programmes. Personally, I think that the imparting of integrated logistic techniques is the most important point in this connection. Even in a transport-oriented country such as the Netherlands, it was recently stated at a symposium that "...integrated logistics are for most Dutch firms but a tempting perspective today" (Van Rens, December 1987). It was added that there would be room in Europe for only 10 to 15 of these integrated transportation companies, acting in the distribution sector.

b) For the middle management of road transport firms, we can adopt more or less the same professional profile as for the heads of the firms. Eventually, in any case, the best of them should become the heads of firms. In the EEC countries, the certificate of competence for access to the profession of road transporter is, in all respects, only the minimum requirement, next to a secondary or university training in applied economics, law or technical specialist knowledge, according to the particular authority to be exercised.

c) Likewise, a new professional profile will be drawn for the drivers of commercial vehicles. For this category, it will no longer suffice in the future to be able to drive and manoeuvre heavy road vehicles, tractors and trailers. In addition they must know how to cope with the electronic equipment of their vehicles and to act as an extension of the central business management, a link in the information system. The drivers will assume commercial responsibilities. In certain circumstances, they will assume legally valid commitments for their firm by their signature or their declarations. During the transportation of dangerous substances, their role in preventing accidents or in limiting the consequences of accidents is often of a decisive nature. Even more than in the past, the drivers will be required to have knowledge of local circumstances and administrative regulations, including some knowledge of languages for road transportation in the open European market where, to a certain extent, inland cabotage traffic is permitted to non-residential transporters.

The training of professional drivers needs to be adapted to this broader profile. In various EEC countries, according to an EEC Directive, there is already special training organized for professional drivers that entitles then to a corresponding certificate of professional competence, apart from the training that leads to a licence for heavy vehicles. In the context of the ADR regulation, an additional level of training is mandatory for drivers of tanker vehicles transporting dangerous substances in fluid, gaseous or solid form. Proposals have been tabled or are being examined to make additional training mandatory for other categories of transportation of dangerous substances, such as dangerous goods in drums or in parcels.

Specific professional requirements have also been set for other special categories or drivers, such as in the subsectors of perishible goods and removals. All of this also naturally applies also to drivers employed in firms with own account-transport.

2. The shipping business and own-account-transportation is a second important category in the demand side for transport professionals. It concerns the traffic managers in industry, the heads of the dispatch or distribution departments in commercial businesses, the manager of the vehicle fleet in firms with much transport on their own account. It is seldom that precise professional profiles have been drawn. In many cases, recruitment is made from within the middle management of the own business, or people are taken

over from the middle management of the professional goods transport or of the forwarding business.

Future needs will be focused on people who are able to build up or use an integrated logistic system and an information system together with the transport business.

3. Forwarding brokers have always played an important role in goods transportation as intermediaries between shippers and carriers. In the new trend towards integrated logistic systems, this task should create a demand for "transportation organizers" as they come to the market looking for trained "all-round" professionals of transportation. A new profile of a "forwarder" is announced, with less "Selbsteintritt", being more of an organizer of the logistic chain rather than participating in the transportation with company vehicles. In this respect an analysis in Germany indicates that the share of staff costs in the forwarding industry increased from 48% to 52% between 1982 and 1987. One can conclude that this happened because higher qualified and therefore higher paid staff personnel were recruited as the need for it was strongly felt. Qualifications are wanted and therefore must be prepared.

4. Providers of services, suppliers and supply companies have just as great a need of people with knowledge of transportation or who have trained in particular aspects thereof.

   Here I am thinking of insurance and credit services, software designers and the like. Above all, the large sector of vehicle constructors needs engineers and designers with a broad knowledge of the transportation sector for which they work. Also, the marketing men and salesmen of the construction firms must naturally have a good knowledge of the world of transportation and be able to anticipate future developments.

   This also applies to consultants and business advisors who want to counsel road transporters and help them with business management.

5. Government departments and professional associations, both national and international, can also be seen as potential employers for personnel trained in the transportation business. The government decides on the infrastructure works that must respond to the transportation needs of the future, and it also deals with the regulations governing transportation. The professional organizations defend and further the interest of transportation. Here, knowledge of the business is an indispensable condition for involvement.

6. The education sector itself, and various training bodies also have demands for transportation experts. Here they mostly call on people who have already had the opportunity of gaining practical experience. Alternatively, the teachers in the training bodies are given the opportunity of going through practical training periods in transportation businesses.

15.5 THE SUPPLY SIDE

According to the above, it seems that transport education and training are going to be the subject of increasing attention in the general education and vocational training systems. People are becoming increasingly aware of transportation, both as regards its economic impact and the external effects of transport in respect of traffic safety and environmental management. There is a desire for transport to be competently organized and carried out, and thus an expectation that people be adequately trained for it. Yet the supply in the field of transport education and training is very disparate and varies dramatically from one country to another. As far as I know, there is little or no comparative literature on this matter, and much research remains to be done in this connection.

It is thus impossible to present the supply side in a generally valid manner. At the most, a few indications and trends can be noted.

1. At many European universities, polytechnics and technical colleges, transportation is not taught at all as such, or at best it is introduced by the back door in one or other of its aspects. Transportation law, insofar as it is addressed in some university faculties, is generally confined to maritime law and air transportation law. In economics faculties, the course is sometimes limited to certain modes of transport and the organization of public transport. There is seldom any complete account of road transportation, for example.
   Yet, we can increasingly see the emergence of institutes and specialist departments of universities for scientific research and education in transportation. According to the universities, and the competence of the professors involved, the accent is variously placed: on technical aspects, on territorial planning and spatial development, on the logistic approach, etc.
   In some countries, particular initiatives have been taken to arrive at some degree of coordination or at least to exchanges of experience. I would cite in this connection the Deutsche Verkehrswissenschaftliche Gesellschaft (DVWG) and, at a European level, the European Society of Transport Institutes (ESTI). Another successful example is the Benelux Interuniversitaire Groepering wan Vervoerseconomen (BIVEC) which groups together professors in transportation economics and related subjects in all Dutch and Belgian universities, as well as their assistants and, as collective members, the various Belgian, Dutch and Luxembourg institutes active in the field of transport.
2. At the level of general secondary education, there are even fewer courses concerning transportation as a specific subject, as far as I know. Naturally, here too the situation must be very different from one country to another, and in various countries, there must be some secondary schools offering courses in transport, shipping and customs systems, for example. I think the trend is

moving towards a sort of integration of subjects connected with foreign trade, logistics and information technology.

3 In various countries it seems to be particularly specialist institutes or similar institutions which are responsible for education and training in transport. Course packages that correspond to the needs at various levels ar offered:

- higher education in transportation science
- basic courses for access to the profession: for future road transporters or harbour employees, for example
- basic technical training for professional drivers, particularly in the transport of dangerous substances or perishable products
- mid-career training in the context of refresher training
- higher training for candidate managers, leading to the grant of the corresponding specific diploma or certificate.

As I said, organization and presentation varies from country to country. It is probably neither possible nor desirable to make all of this uniform. The only thing we need to consider, especially in Europe, is whether the supply of transport education and training is sufficiently attuned to the future needs of the sector.

4. I would also like to make a special note of the supply of refresher training that is on the market in various forms, sometimes in over-supply. These are study days, refresher training courses, seminars, colloquia, conferences, etc... The starting point is the justified assumption that the business world, in its technical, financial and commercial aspects - but, above all, as a consequence of information technology - is going through radical changes and developing faster than ever before. This leads to the need for continuing training, refresher training and adult education.

The range of the supply of subjects in the field of transportation, includes post-academic education via programmes in the context of training in small and medium firms, initiation to new techniques, updating of knowledge, and recycling of certain specialist trainings. In this connection, training and refresher training is also organized within firms, with or without help from outside, with programmes that are tailor made to meet the needs of the particular firm.

15.6 CONCLUSIONS

1. It is established that the transportation of goods, in particular transportation by road, will go through a series of rapid and radical changes, probably in Europe in particular. The professionals of transportation, at all levels, will

have to approach transportation in a different manner, taking technical and technological progress and new market situations into account. Thus new additional knowledge and skills will be demanded of the transportation professionals. The present generation is not trained for this, or inadequately trained. Thus education, training, and refresher training must be responsible for the preparation for the future.

2. Transportation and shipping firms, together with all of the business sectors involved, will have to determine what sort of transportation professionals they need in the future and what their qualifications need to be. This demand side must be determined in good time by enquiries and research. At the same time, the quantitative needs must be determined six to ten years ahead. In the European countries, this research should be embarked on urgently, if they are to be able to attune the supply of education to future needs in good time.

3 In the meantime, the supply side of transport education, training and refresher training needs to be made more transparent and, where applicable, better coordinated. In most European countries it is difficult to get a clear picture of where transport education and training is currently on offer on the market.

4. The future standards for transportation professionals must be translated into education procedures, so that training curricula and programmes can be worked out or adapted accordingly.

5. Without wishing to strive for a uniform system, international exchanges of experience and cooperation would be eminently desirable. International organizations such as the EEC Commission, the European Conference of Transport Ministers, the International Road Transport Union (IRU) and the like can contribute to this. As can a conference such as today's.

REFERENCES

Alaluf, M., a.o. (1988). *Adéquation Enseignement - Marché de l'Emploi.* Brussels, Belgium: Fondation Roi Baudouin. ISBN: 21-872-005-X.

Beagley, T. L. (1988). Training and Education in Transport: the European Scene. Brussels, Belgium: European Society of Transport Institutes (E.S.T.I.).

De Waele, A. and Violland, M. (1987). *Research relevant to trends in transport over the coming decade.* Paris, France: European Conference of Ministers of Transport (E.C.M.T.), Round Table 75. ISBN: 92-821-1118-0.

Dewaele, A., a.o. (1988). *Aansluiting Onderwijs-Arbeidsmarkt.* Brussels, Belgium: Koning Boudewijnstichting. ISBN: 90-5130-014-X.

Jaumotte, A. L. (1988). *Réflexions sur la dynamique des sociétés industrielles et sur les transports routiers de demain.* Paris, France: OECD, Colloquim preprints.

Khisty, C. J., a.o. (1986). *Transportations and Training.* Washington, DC: Transportation Research Board, Record 1101.

Meyer, M. D., o.a. (1985). *Transportation Education and Training:* Meeting the Challenge. Washington, DC: National Research Board, Report 210. ISBN: 0-309-03914-2.

Soukkha, H. M., a.o. (1988). *Telematics-Transportation and Spatial Development.* Delft, The Netherlands: Centre for transportation engineering (Delft University of Technology), Symposium preprints.

Van Rens, J. H. P., o.a. (1987). *Freight Logistics and Information Technology.* The Hague, Netherlands: Hague Consulting Group, Symposium preprints.

Van Stappen, J. (1987). *La condition de capacité professionnelle exigée des transporteurs routiers de la C.E.E.: L'application actuelle et le développement futur.* Brussels, Belgium: European Society of Transport Institutes, Seminar on Transport Training and Education (Lyon, France), preprints.

Chapter 16

THE INFRASTRUCTURAL CHALLENGE FOR FUTURE TRANSPORTATION - SOME CRITICAL ISSUES

Dan Näsman,
Swedish National Road Administration

16.1 INTRODUCTION

In almost every industrial investment decision the infrastructure of the transport sector is taken for granted. The same goes for the government's long term economic planning and most of the policymaking in fields like social welfare, education, regional policy and employment. And rightly so, when we take into account the very long run that applies to the planning and construction of such basic facilities as railways, airports, road networks etc. The investment needs that we identify today will normally take between ten and fifteen years of planning, decisionmaking and construction to complete (if they are possible at all). Then we will use these constructions for 30-40 years (or perhaps even longer). We need to make the right investments, because the expenses will be so high that we will not be able to afford to correct any mistakes within a reasonable period of time. The conclusion is that what we are planning for today will to an important extent define our transport conditions for the next fifty years or so.

Why do I stress this point? I realize that what I have said so far is evident to all of you. Nevertheless it seems to be an inevitable fact that problems and shortcomings of the transport system of today - which are the consequenses of decisions taken decades ago - are dominating the public debate and political discussions to an extent that almost totally supresses considerations of the long run tendencies and needs of tomorrow.

From this point of view I think it would be justifiable to try to point out some questions that seem to be of strategic importance in order to focus some attention upon the needs for transportation in the future.

To illustrate the situation I will briefly describe a few dominant and - as I presume - long run tendencies in the needs for transport services.

Then I will roughly characterise the framework in which the investments for transport purposes is taking place.

My examples will be from Sweden. I am convinced, however, that the general tendencies apply to most of the modern industrial countries. And, being a representative of the National Road Administration, I hope you will forgive if I focus on matters concerning the road network.

16.2 SOME TENDENCIES IN TRANSPORTATION DEVELOPMENT

Starting with passenger transport the first thing to be aware of is the dominant role of the road and street network. 85 percent of all forms of transport take place on roads and streets. Of the remaining 15 percent, 12 percent take place in subways and railways and 3 percent by air.

Is there anything in the foreseeable future that could reduce the importance of the road network? Well, as far as I can see, the demands and tendencies are going in the opposite direction. This view is based on a number of basic tendencies in Swedish society.

First, there is a very clear tendency in the ownership and use of private cars. From a situation when one car was the natural aim for one household there is now a gradual shift towards individual car ownership. This is due to basic changes in the functioning of a household related to labour market conditions, the tax system, the upbringing and education of children and youths etc. Together these circumstances require each individual to be highly mobile on a daily basis.

Second, we observe the increasing number of retired people who have driving licences and cars of their own. This is something new compared to the situation just a decade ago and it certainly will result in a substantial increase in the demand for road transportation.

Third, there is a very strong and determinate public and political opinion that the population of the country, and I suppose most countries, should not be confined to living in a few very densely populated urban areas with the rest of the country being more or less deserted. On the contrary, large efforts are being made to support and stimulate industrial life and living conditions in areas that would otherwise be in danger of decline. The only way to promote necessary personal mobility in these areas is by road transport.

There is, however, one area in which a shift from road transportation to other facilities seems to be possible and necessary. I refer of course to big city areas in which the road traffic situation already is, or is quickly approaching, an unacceptable situation.

But the overall conclusion is very clear: There is nothing which currently indicates that the role of road transportation will be less important in the future than it is today as long as we do not want to heavily restrict the mobility of some part of the populationand as long as we accept that the population should not be concentrated to a number of geographically limited areas. Another way to put it is to say that the demand for road transport follows directly from fundamental political goals in the society.

When it comes to transportation of goods the role of the different means of transportation varies between different industrial countries depending on differences in their industrial structure, the structure of foreign trade, physical distance between production sites and markets etc. In Sweden, the transportation of goods is divided between roads, railways and ships; they represent 53, 25 and 22 percent respectively. Although road transport is not as dominant as it is in the case of passenger transport, it nevertheless stands for more than half of the transported goods.

What about future developments in this area? Well, you are probably all familiar with the gradual shift in most industrial countries towards the increasing importance of manufacturing and services industries, often at the expense of traditional raw material-based industries. In the Swedish case this is clearly demonstrated by this figure (Figure 16.1, Andersson and Strömquist, 1988).

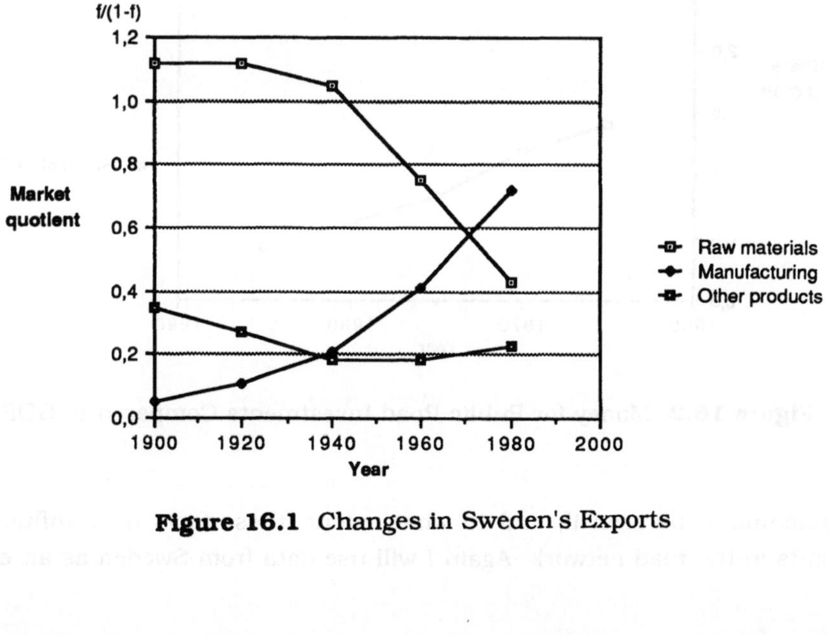

Figure 16.1 Changes in Sweden's Exports

It is probably sufficient to merely mention the concept of just-in-time and the necessity to reduce capital costs in manufacturing etc. to make it clear what this shift means in terms of demands for more frequent, reliable and flexible goods transport.

The conclusion is clear once again: There is nothing that indicates that the transportation of goods on roads and streets will be of less importance in the future than today. On the contrary the basic economic forces behind the tendencies I have just indicated tend to shift the flows of goods from bulk transport by sea and railway to roads and, in some cases, to air transports.

16.3 INVESTMENTS IN THE ROAD NETWORK

So far I hope I have established two facts; one being the necessity of long term planning for transport infrastructure and the second the conclusion that the importance of a road network adapted to its purpose will be more rather than less important in the future if the mobility of all parts of the population and the competitiveness of the industries are not to be jeopardized.

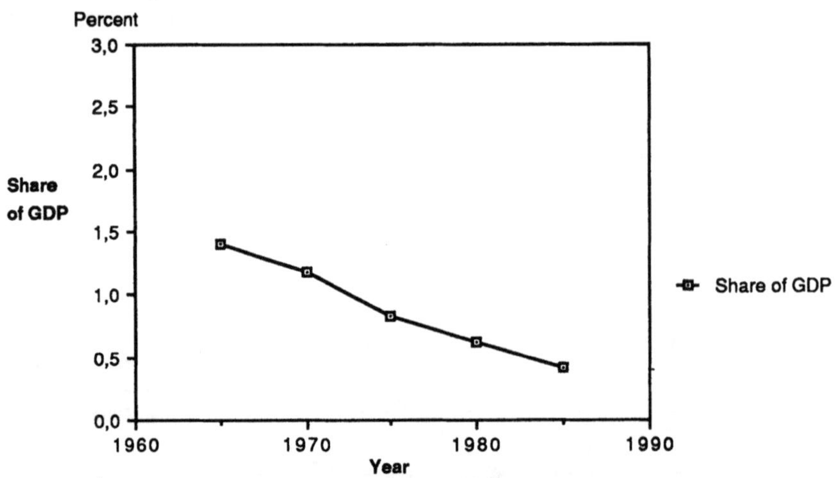

Figure 16.2 Money for Public Road Investments Compared to GDP

Now it remains to be seen if, and to what extent, these facts have influenced the investments in the road network. Again I will use data from Sweden as an example.

One quantitative approach to that question is to consider the share of the gross domestic product that is used for investments in the road network.

Figure 16.2 gives a clear answer. In the 20-year period from 1965 to 1985 the share has been reduced from about 1.5 percent to less than 0.5 percent of the GDP. Certainly, this is a worrying fact in relation to the long run tendencies for road transports. Will it really be possible to meet the growing demands for high-quality transports that the future calls for?

Let us go one step further and ask an even more fundamental question. Is it possible that we will be able to make the necessary investments to replace those parts of the road network which are worn out? Maybe "worn out" is not the proper expression for what I'm trying to say. It is not only a question of physical decay but also a matter of economic obsolescence due to the changes in the flows of goods and people. Let us assume that the normal lifetime of a road is 40 years due to the combined effects of physical and economical factors. If this is so, then Figures 16.3 a and b give a clear and frightening picture of future road transports in Sweden.

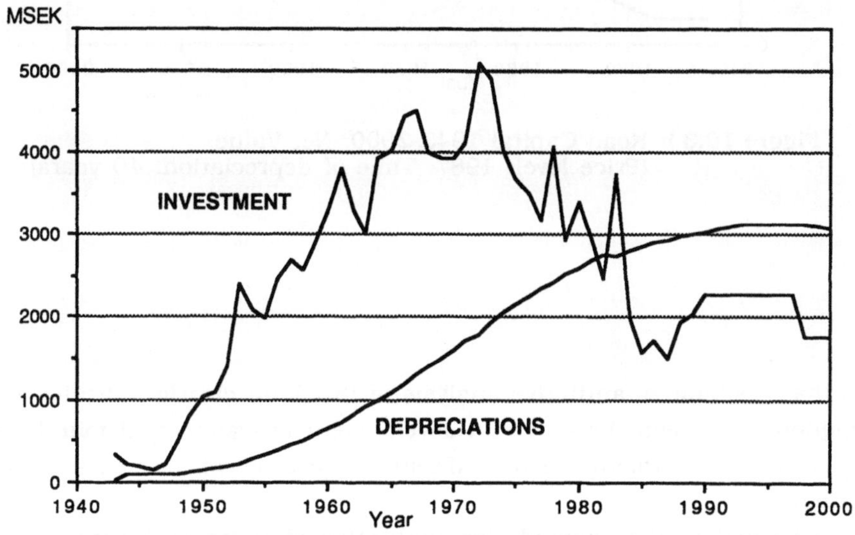

Figure 16.3 a Annual Road Investment and Depreciations 1943-2000. (Price level: 1987. Time of depreciation: 40 years)

To maintain the road network standard of today we have to invest 1/40 of all previous investments each year in order to compensate for the depreciation of that year. As can be seen from the figures, the value of the Swedish state road network was increasing until the early 1980s - as long as the level of investment was larger than the depreciations. But since then the level of investment has been too low to preserve the value of the previously invested capital. The increases that have been announced in recent political decisions will not be sufficient to change the direction of this development.

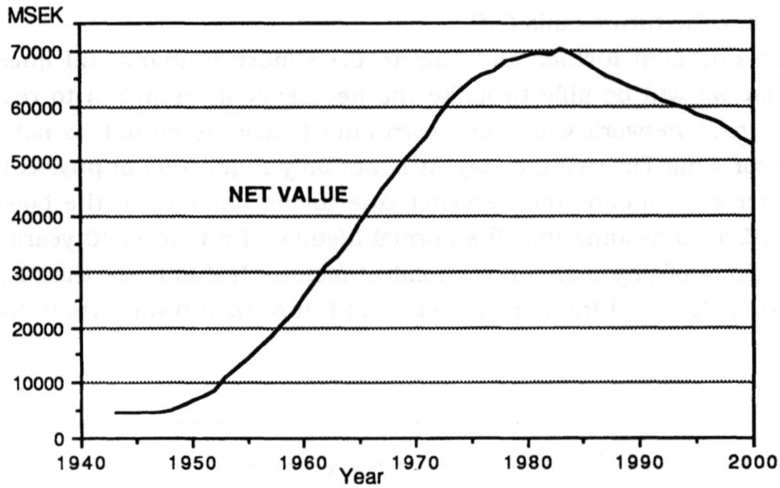

Figure 16.3 b Road Capital 1943-2000, Net Value. (Price level: 1987. Time of depreciation: 40 years)

It seems that a simple quantitative analysis of this kind clearly indicates a lack of correspondence between the long run development of transport demands and the planning for and investments in the infrastructure that will be necessary to meet the demands.

If we then consider the demands for qualitative changes in the transports ,it is evident that the shortcomings of the infrastructure will be even more serious in the future.

Furthermore, the demands for investments in infrastructure are not only a matter of quality of transportation. As we all know that road transportation causes serious damage to the environment, affecting both human and ecological life. Satisfactory solutions to these problems are certainly a condition for future road transportation (as well as for industrial life, I may add). The most fundamental and important solutions are probably a matter of technological changes in the vehicles themselves. But it is also evident that major undertakings will be necessary to adapt the infrastructure to environmental demands in at least the large city areas.

16.4 SOME CRITICAL ISSUES FOR THE FUTURE

The very long run cycles that apply to investments in the infrastructure of transportation and the apparent lack of correspondence between future demands and the investment activities of today raisessome urgent questions.

First. I think it is fair to describe the current planning of infrastructure as adaptive or passive in the sense that almost all efforts are spent in solving or reducing the transport problems that are acute at present. Typical examples are bottleneck-problems in the road network, noise or pollution problems in large urban areas, the replacement of bridges that cannot carry heavy trucks, etc. In view of this it seems natural to ask for a more anticipatory or visionary direction of the investments. Is it possible to break the vicious circle of dealing with the problems that were created yesterday to day and replace it with a benevolent circle in which we invest today to avoid the problems of tomorrow?

Do we have the necessary knowledge and planning methods to do it? Could it be done in such a convincing way that will be necessary to bring public consciousness and political support?

Second. turning back to the quantitative aspect of the investments; will the institutional and financial conditions of today be sufficient to provide the necessary financial resources? Although there evidently is an increasing demand for high-quality road transportation in most countries there is no direct financial link between the taxes on road traffic and the public expenditures on road investments. Therefore, it seems like the general tax level and the fiscal policy of a country becomes an obstacle to the working of the market forces in the transportation sector. Figure 16.4 illustrates this argument.

The costs to build, maintain and operate the road network are just a small part of the total costs for road transportation (far bigger parts being vehicle costs, travel time and road accidents). There is strong evidence that the total costs could be reduced if an increased share was used on the road network. How are we to find a politically acceptable method to transfer the necessary investments funds to this sector?

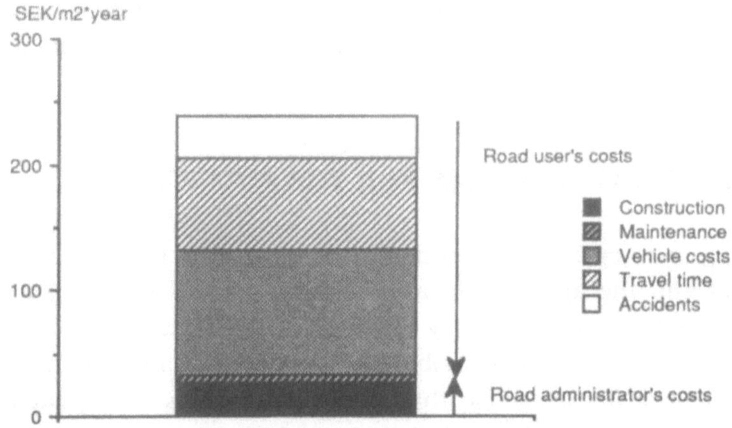

Figure 16.4 Socioeconomic Costs for a 7 m Wide Road

Third, will we be increasingly troubled by capacity limits on the road network? Today we already see that kind of problem in some areas. If the tendencies in future road transportation and road investments that I have described apply to other, more densely populated countries than Sweden I think this is a well justified question. At the same time we are aware of the fact that the road network is very unevenly utilized during different hours of the day. Therefore, isn't it likely that in the near future it will be necessary to introduce some kind of regulations in order to distribute the traffic during the hours of the day? If so, what impact will this have in different areas of working life and social life?

Fourth, and finally, there is cause to question what the impact of new information technologies and other major technological break-throughs of the transportation sector will be. Could it be that technological changes in vehicles and their means of navigation alters the demands for developments of the infrastructure? Is it even possible that the risks of serious shortcomings of infrastructure which follow from my line of reasoning are exaggerated as a result of technological possibilities which are unknown to me?

When I put these questions I am aware of the fact that they might appear to be relevant to a different degree in different countries. I believe, however, that the differences between countries in this case is more a matter of how far the situation has advanced than a matter of fundamental differences. And that is the reason why I venture to say that these problems are urgent and should be dealt with before it's too late. The future never just happens, it is created.

REFERENCES

Andersson, Å. E. and Strömquist, U. (1988). *K-Samhällets Framtid*. Sweden: Prisma. (In Swedish).

Appendix

PROMETHEUS: THE EUROPEAN PROGRAM FOR IMPROVED ROAD TRAFFIC

Tage Karlsson
International Steering Committee of Prometheus

In October 1986, the indigenous European automotive companies initiated an eight year cooperative research and action program between themselves, their suppliers, universities and other research institutions, authorities and European governments. The objectives of this program are to improve safety, increase efficiency and lessen the automotive effect on the environment. As far as safety and the environmental effect are concerned, a 50% improvement is sought by the start of the new millenium. The efficiency improvements - including the use of roads, utilization of time, energy savings, stimulation of research, industry and infrastructure construction - are so manifold that a satisfactory quantification has not yet been defined, but rough calculations indicate great savings for the European people.

Based on well founded analyses of the future needs of their products, the automotive companies in the sixties did not heed the signs when the drive for greater safety and a better environment started. Thus, with some exceptions, they gained a reputation for obstructing efforts to improve these quality of life components. The Prometheus program is a definite signal that the European automotive companies are very willing to cooperate with all parties concerned for a safe, clean and more efficient utilization of automotive mobility.

But Prometheus is not only founded on a change of mind regarding the social responsibilities of companies for their products and for the use of their products. The main basis for Prometheus is the rapid progress of microelectronics and information technologies. Advances in these areas will have considerable effects on automotive vehicles. Some forecasts say that electronic products at the turn of the century will double to about 25% of the value of the cars compared to the situation today. The effect of this doubling is further magnified because of the very rapid decrease in the cost of such products.

The new electronic products can be installed as devices, one at a time, using trial and error to gradually adapt the new devices to each other. This method will produce different solutions from different manufacturers, and thus harmonized systems in and between European countries will not eventuate within the foreseeable future.

The Prometheus approach is the systems approach: trying to find the best system combinations of (1) microelectronics, (2) information technology, (3) road traffic and (4) infrastructure for roads and telecommunications - in order to improve safety, efficiency and the environment. To find the optimal solutions, the European automotive industry is cooperating in precompetitive research and the intention is that many of the innovations shall be introduced as industry standards. But that is only possible for a relatively small share of the total mass of innovations. Immediate automatic communication between vehicles and the roadside, communication centers and satellites requires complete cooperation from the telecommunication authorities and the road authorities. One problem is to find and define the methods. Another is to adapt the infrastructure for them and that can only be done by governments and traffic authorities. If this is to be done on a Pan European basis, we need government agreements, laws, standards and protocols. Western Europe is governed by 28 sovereign governments. Even if the automotive companies succeed in obtaining the full cooperation from their own five host countries, they will need the full support of these governments to convince the other twenty-three.

Prometheus needs to be in the very frontline of research into electronics and information technology, i.e. in areas which do not properly belong to the automotive industry. Therefore the best scientists from the leading research institutions in Europe have been invited to do basic research for Prometheus. The supplier industries have been asked to develop prototype proposals and to do applied research within their respective component areas.

The organizational structure of Prometheus is led by the International Steering Committee of 11 members, each one representing one or more of the participating automotive companies, namely: Daimler Benz, Volkswagen, BMW and Porsche from West Germany; Renault, Paugeot and Matra from France; Rover, Rolls Royce and Jaguar from Great Britain; Fiat and Alfa Romeo from Italy; and Saab Scania and Volvo from Sweden. In parallel with the International Steering Committee is the Pro-Council, made up of representatives from the host countries as a coordinating link with the governments and authorities.

The research performed by the automotive industry is arranged in 11 thematic groups which treat subjects of common interest, and in 11 company demonstrator programs where each automotive company contributes to the Prometheus program. There are four international basic research groups dealing with Artificial

Intelligence, Computer Hardware, Communication Techniques and Road Infrastructure.

Members of the International Steering Committee and their deputies serve in addition as links to the Thematic groups and to the Basic Research groups. Each of these latter groups have their own international coordinators. There is also a central coordinating office which arranges meetings, makes the protocols and handles the arhives of the project. Once the main guidelines have been established and Prometheus settles into a regular working pattern, the intention is to convert the International Steering Committee into a supervisory board with the coordinating office promoted to perform an executive function. The Steering Committee has a subcommittee called the Impact Analysis Committee, charged with the task to study the intended, expected and realized impact of Prometheus on safety, efficiency, the environment and society in general.

The Prometheus project is planned to continue for 8 years. The first year was used to describe the state of the art of the different knowledge components of Prometheus, and from there to specify the necessary research topics. The second year has been allocated to recruit the help of suppliers and to launch the specified research tasks. For the moment, some 400 engineers and scientists are taking part in the work of Prometheus.

The remaining 6 years to the end of 1994 have not yet been planned in detail, but there is little time to lose if the stated objectives are to be achieved. Rough estimates of the cost of the research program show a total of some 600 million ECU (i.e. some 725 million USD), but these calculations are quite uncertain. What is expected is that the implementation of the resulting benefits of the research program - in the form of new components, new systems, modified road and communication infrastructures - will dwarf the costs of the research program. This should not be looked upon simply as a cost but rather as an investment in development, growth and employment, which can benefit all of Europe.

It is expected that the Prometheus research program will stimulate European research in the relevant areas to levels comparable with those of the USA and Japan, and that the industrial implementation will make the European manufacturers competitive with their overseas rivals. Prometheus is a European project, intended to solve European problems with the help of European financing. Because of this latter aspect there are some restrictions on participation in the program. It is of prime importance to secure the cooperation of the European governments, universities, transport authorities and industry. This is a gigantic task in itself which has to be prioritized before realization. When that is well on its way, Prometheus will consider cooperation with non-European organizations.

List of Contributors

Prof. Åke E. Andersson	Institute for Futures Studies Hagagatan 23 A S-113 47 Stockholm SWEDEN
Prof. David F. Batten	Centre for Regional Science Research (CERUM) and Department of Economics University of Umeå S-901 87 Umeå SWEDEN
Prof. David E. Boyce	Urban Transportation Centre (M-C 357) Suite 700 South 1033 Westwanburen Street Chicago, Illinois 60607-9940 U.S.A.
Dr. George Eads	Vice President and Chief Economist General Motors Corporation 3044 West Grand Boulevard Detroit, Michigan 48202 U.S.A.
Mr. Stig Ericsson	Vice President Saab-Scania AB, Scania Division S-151 87 Södertälje SWEDEN
Mr. Robert L. French	R. L. French & Associates 3815 Lisbon St, Suite 201 Fort Worth. Tx 76107 U.S.A.
Dr. Haruki Fujii	Manager, Research Department Association of Electronic Technology for Automobile Traffic and Driving (JSK) Toranomon -34 Mori Building 1-25-5 Toranomon, Minato-Ku Tokyo 105 JAPAN
Prof. William Garrison	Department of Civil Engineering Institute of Transportation Studies University of California Berkeley CA 94720 U.S.A.
Dr. Johannes Grevsmähl	Bundesministerium für Verkehr Postfach 20 01 00 D-5300 Bonn 2 FEDERAL REPUBLIC OF GERMANY

Mr. Tage Karlsson	Project Manager AB Volvo Dept 6090 S-405 08 Göteborg SWEDEN
Prof. Dr. Ing. Karl Krell	Head of Traffic Engineering Bundesanstalt für Strassenwesen Brüderstrasse 53 D-5060 Bergisch Gladbach 1 FEDERAL REPUBLIC OF GERMANY
Dr. Nebojsa Nakicenovic	Technology, Economy & Society Program International Institute for Applied Systems Analysis (IIASA) A-2361 Laxenburg AUSTRIA
Dr. Curt Nicolin	Chairman ASEA Box 7373 S-103 91 Stockholm SWEDEN
Mr. Dan Näsman	Finance Manager Swedish National Road Administration S-781 87 Borlänge SWEDEN
Prof. Dr. Hellmuth St. Seidenfus	Director Institute of Transport Economics University of Münster FEDERAL REPUBLIC OF GERMANY
Dr Joseph van Stappen	Conseiller Général Institut du Transport Routier Rue Archimede 5 B-1040 Bruxelles BELGIUM
Mr. Ulf Strömquist	General Manager TEMAPLAN AB Sandhamnsgatan 81 S-115 28 Stockholm SWEDEN
Prof. Carl-Olof Ternryd	Chairman The Association of Swedish Automobile Manufactures and Wholesalers Box 5514 S-114 85 Stockholm SWEDEN
Dr. Roland Thord	TEMAPLAN AB Sandhamnsgatan 81 S-115 28 Stockholm SWEDEN

MIX
Papier aus verantwortungsvollen Quellen
Paper from responsible sources
FSC® C105338

If you have any concerns about our products,
you can contact us on
ProductSafety@springernature.com

In case Publisher is established outside the EU,
the EU authorized representative is:
Springer Nature Customer Service Center GmbH
Europaplatz 3, 69115 Heidelberg, Germany

Printed by Libri Plureos GmbH
in Hamburg, Germany